SpringerBriefs in Applied Sciences and Technology

SpringerBriefs present concise summaries of cutting-edge research and practical applications across a wide spectrum of fields. Featuring compact volumes of 50 to 125 pages, the series covers a range of content from professional to academic.

Typical publications can be:

- A timely report of state-of-the art methods
- An introduction to or a manual for the application of mathematical or computer techniques
- A bridge between new research results, as published in journal articles
- A snapshot of a hot or emerging topic
- An in-depth case study
- A presentation of core concepts that students must understand in order to make independent contributions

SpringerBriefs are characterized by fast, global electronic dissemination, standard publishing contracts, standardized manuscript preparation and formatting guidelines, and expedited production schedules.

On the one hand, **SpringerBriefs in Applied Sciences and Technology** are devoted to the publication of fundamentals and applications within the different classical engineering disciplines as well as in interdisciplinary fields that recently emerged between these areas. On the other hand, as the boundary separating fundamental research and applied technology is more and more dissolving, this series is particularly open to trans-disciplinary topics between fundamental science and engineering.

Indexed by EI-Compendex, SCOPUS and Springerlink.

Akash Abaji Kadam

AI-Powered Smart Manufacturing

Practical Decision-Intelligence Frameworks for Small and Medium Enterprises

Akash Abaji Kadam
Mechanical Engineer
Servotech Inc
Chicago, IL, USA

ISSN 2191-530X ISSN 2191-5318 (electronic)
SpringerBriefs in Applied Sciences and Technology
ISBN 978-3-032-21465-2 ISBN 978-3-032-21466-9 (eBook)
https://doi.org/10.1007/978-3-032-21466-9

This Springer imprint is published by the registered company Springer Nature Switzerland AG
The registered company address is: Gewerbestrasse 11, 6330 Cham, Switzerland

Preface

Small and medium enterprises (SMEs) operate in a world of manufacturing that is constantly under pressure. It starts with changes to the manufacturing schedule every day, the warehousing of products that ties down funds, machine breakdowns, and supplier uncertainties that cascade into all other areas. Compared to larger companies, SMEs lack resources and staff to deal with all this.

This book speaks of a very basic yet commonly overlooked concept: most of a manufacturer's issues do not relate to technology but to decisions. Small and mid-size businesses do not collapse because of a lack of AI technologies; they struggle because they do not have clear access to information that can help people make better decisions.

This particular book is not about training oneself in AI or math for that matter. Instead, it provides insight into how all those components from AI to analytics and digital tools can be combined effectively for developing decision intelligence solutions. Decision intelligence solutions assist people in making smarter decisions at the production level or at the planning level and at the procurement level.

This book is for:

- Manufacturing engineers and supervisors
- Production planners and inventory managers
- Maintenance and reliability professionals
- Owners and managers of SMEs
- Graduate students searching for industry-oriented concepts

In the chapters, difficult concepts will be illustrated using examples based on manufacturing, and through step-by-step processes which can be applied by small and medium-sized enterprises. The people-first strategy, in keeping with Industry 5.0, has always had AI as an aid to, not a substitute for, this strategy.

The reader does not have to be an expert in data science or machine learning. It is important that they understand manufacturing and be willing to change the process of decision making when faced with uncertainty.

Chicago, IL, USA Akash Abaji Kadam

Acknowledgments

The present book is the result of numerous years of experience working with manufacturing professionals and dealing with everyday operating issues. The author thanks the engineers and managers whose experience has contributed to implementing the ideas conveyed in this book.

The author would like to extend special acknowledgement to some reviewers and colleagues whose comments and contributions toward the clarification and usefulness of the topic were invaluable. The author is grateful to the manufacturing and research communities whose efforts in smart manufacturing, Industry 4.0, and Industry 5.0 continue to push the boundaries.

Lastly, special thanks go to the family for their patience and support during the process of working on this thesis.

Competing Interests The author has no competing interests to declare that are relevant to the content of this manuscript.

Contents

About the Author

Akash Kadam is an experienced mechanical engineer and also an independent researcher in smart supply chain systems, advanced manufacturing, and design innovation. He is an M.E. in Mechanical Engineering from Texas A&M University–Kingsville and has over a decade of professional experience. He has led influential, self-driven research summing up AI, IoT, blockchain, and sustainability in Industry 4.0 and 5.0 paradigms. Through multiple authored papers and strategic outlooks, he contributes to operational resilience, digital transformation, and quality excellence. Through his work, it is demonstrated that the future of smart manufacturing and global supply chain leadership encompasses substantial original contributions.

Chapter 1
Introduction to Smart Manufacturing and Industry 4.0–5.0

Introduction Manufacturing has long been more than churning out goods; it is a backbone of economies, of how people make a living, and of technological development. Factories have been the human development engine for centuries, determining how societies develop and prosper. But something extraordinary has occurred in the past decade: manufacturing began to think, learn, and adapt. No longer do machines merely do; machines now perceive, interpret, and make better.

This evolution is what we refer to as smart manufacturing. It is a transition from mechanical only to cognitive, data-based environments that bring human and technology together. For giant companies, embracing such technologies has been easy due to their finances and availability of skills. But this is more difficult and more worthwhile for small and medium companies (SMEs). SMEs are most of all manufacturers of the world, and equipping them with smart solutions could reorient how industries will compete in the years to come.

We will delve into how manufacturing itself has transformed through various epochs, grasp how smart manufacturing really is, and get a glimpse of how a transition from Industry 4.0 to Industry 5.0 is affecting SMEs. We will conclude with a real-world preparedness checklist so that SMEs have a baseline against which to measure themselves before delving further into the book.

1.1 Evolution of Manufacturing: From Mechanization to Digitization

Manufacturing is a story of revolutions—new periods introducing innovations that transformed production as much as life and work itself. Comprehending this historical development enables SMEs to appreciate that shrewd production is no surprise shock but is instead a continuation of a long tradition.

A. A. Kadam, *AI-Powered Smart Manufacturing*, SpringerBriefs in Applied Sciences and Technology, https://doi.org/10.1007/978-3-032-21466-9_1

The First Industrial Revolution: During the closing years of the eighteenth century, factories saw manual labor displaced by steam and water-powered machinery. Textile mills, blast furnaces, and mechanical looms converted small shops into giant production facilities. Manpower no longer set a limit to productivity.

Effects on SMEs: This period witnessed most of the small workshops growing into small factories. Even small investment in mechanical machines doubled production.

The Second Industrial Revolution (Industry 2.0, Mass Production and Electrification): By the end of the nineteenth and early twentieth centuries, assembly lines and electricity changed manufacturing dramatically. Henry Ford's moving assembly line is, of course, its best-known version, as it allowed mass production of low-cost automobiles. Steel production, chemicals, and machine-building businesses thrived.

SME Impact: Electrification made it possible for even tiny producers to raise efficiency. SMEs embracing powered machines enjoyed huge competitive advantages.

The Third Industrial Revolution (Industry 3.0, Automation and Computers): Since the 1970s, electronics, information systems, and automation recreated factories. Programmable logic controllers (PLCs), computer numerical control (CNC) machines, and primitive robots eliminated repetition work. Factories also started to gather data, even if analysis remained rudimentary.

SME Impact: Automation decreased error rates and improved consistency. High prices, however, saw many SMEs being left behind by bigger businesses.

The Fourth Industrial Revolution (Industry 4.0, Cyber-Physical Systems): Beginning in the 2010s, Industry 4.0 linked machines, sensors, and systems by Internet of Things (IoT), cloud computation, and sophisticated analytics. Factories were made "smart," to anticipate breakdowns, adapt production timetables, and learn by data.

SME Impact: Even though adoption began slowly, SMEs were also accessing affordable IoT devices, cloud platforms, and open-source AI tools. That evened the field, and smaller companies were now free to deploy technologies formerly reserved for large corporations.

The Fifth Industrial Revolution (Industry 5.0, Human-Centric and Sustainable): Today, we are moving into Industry 5.0. Whereas Industry 4.0 placed emphasis on connectivity and automation, Industry 5.0 stresses man-AI cooperation, customization, and sustainability. Instead of displacing workers, man and machine, including robots, serve each other as copartners, supplementing each other's capability and creativity.

SME Impact: It is a silver bullet moment here. SMEs, being agile and having intimate customer contacts, are better positioned to incorporate personalization and sustainability ahead of large companies hindered by bureaucracy.

Key Takeaway for SMEs: Every industrial revolution paid off for adapters early enough. Today, intelligent manufacturing is no longer a choice; it's a doorway to survival and prosperity in a globalizing market. For SMEs, it's both a threat and a challenge to rethink how they add value.

1.2 What Is Smart Manufacturing?

Smart factory is among the greatest revolutions going on in the industrial space nowadays. Basically, it is a new way of thinking about designing, making, and delivering products. It is unlike classical systems of manufacturing, whereby each operates independently, both mechanically and labor-wise, like silos. Smart manufacturing, on the other hand, engulfs technology, people, and processes into a single interlinked system, which is intelligent, connected, and adaptable. It involves more than introducing automation or information systems on the factory floor. It's more about getting the entire manufacturing value chain to operate seamlessly and data-intensively together.

It's helpful to compare smart manufacturing with older ways of doing things to really know what it is. Under traditional manufacturing, production decisions were made based on limited information, experience by man, or past histories. A manager, for instance, might organize machine maintenance once a semester, whether the machine required it or not. Quality tests were usually done by hand, and stock decisions were made by estimates. With this kind of setup, inefficiencies and slowdowns were inevitable, for information traveled slowly and usually got manipulated across departments.

Smart manufacturing shifts this paradigm by setting an atmosphere where information travels in real time across each stage of the value chain. Machines cease being dumb pieces of machinery doing mere routine work. With sensors, controllers, and sophisticated software, machines are capable of creating, gathering, and transmitting information regarding their efficiency. This information may then, instantaneously, be analyzed by machine learning or artificial intelligence programs to give feedback and advice. For instance, rather than closing a machine after a breakdown, predictive analytics may recognize when a machine is likely to fail and inform the operator to undertake repairs before it happens. It is one of the defining aspects of smart manufacturing to act ahead of schedule, rather than being reactive.

Yet another key characteristic of intelligent manufacturing is connectivity. If machines, tools, and production lines are interconnected by the Internet of Things (IoT), data is no longer isolated within a single department or a single production stage. A sensor on a milling machine can talk directly to a central planning system, for example, which then changes scheduling or material demand automatically. This engenders a degree of openness and coordination impossible under older systems. The outcome is not just rapid reaction to disturbances but also a general rise in efficiency.

Intelligence and flexibility are also at the heart of intelligent manufacturing. Unlike rigid assembly lines of the past, intelligent systems learn by observing patterns and adapt their processes to match. A factory for consumer products can change between variations of products rapidly with a minimum of downtime since the system observes the switch and resizes itself. It then becomes possible for even smaller firms to adopt mass customization, delivering custom products to customers without being penalized by high expense or lengthy lead times.

For small- and medium-sized businesses, the concept of smart manufacturing many times seems like it's something for large multinationals with unlimited pockets of money. It is a misconception, though. SMEs actually have a lot to gain by embracing smart things, quite often even more compared to their bigger competitors. A tiny machine shop, for instance, can implement low-cost IoT devices to keep an eye on its CNC machine performance, such as tool wear, energy, and cycle times, for its owner's real-time view. A textile SME can implement artificial intelligence-based scheduling systems to reduce production during peak electricity hours, cutting costs, and lowering its carbon emission profile. A food-processing SME, finally, can establish computer vision systems for automatic defect recognition for package, as well as for quality defect, a thing otherwise requiring a large staff of inspectors, yet still having its inaccuracies.

What is so compelling about these examples is that intelligent manufacturing is not about displacing people but about supplementing their ability. Instead of having to spend hours watching routine work, employees are now able to spend time on higher-order activities like solving, inventing, and servicing customers, and machines handle routine watching and number crunching. It is the symbiotic relationship between people and technology that is so transformative about intelligent manufacturing.

Simply, smart manufacturing refers to a system of changing production by transferring it from guessing to knowing, by changing reaction into prediction, and by running individually to running interconnectedly. It is not merely about running after current trends for SMEs; it is about surviving and being competitive in a global marketplace where efficiency, speed, and quality lead to success. By embracing smart manufacturing, SMEs are able to achieve more by doing less, to base decisions on real fact instead of assumptions, and to future-proof business operations amidst a highly uncertain business landscape.

It's analogous to driving a car by gazing merely at the rear-view mirror: decisions are made based on stuff that's occurred in the past. Smart manufacturing is like having a GPS unit that gives real-time traffic information, identifies where it's going to slow down, and offers advice on how to speed past it quickest. It empowers SMEs to see ahead, act boldly, and stay a step ahead.

1.3 Transition from Industry 4.0 to Industry 5.0

The manufacturing process itself had consistently evolved in waves, with each successive period standing on the shoulders of the prior period's innovations. The Fourth Industrial Revolution, also known as Industry 4.0, became the first to combine seamlessly both worlds, bringing a sense of intelligence, connectivity, and automation to bear on manufacturing production that utterly redrew prior boundaries of what was possible. As revolutionary as it proved, however, even Industry 4.0 showed its limitations. It proved that demonstrable efficiency was possible by driving machines and data, yet it also proved irrefutably true that humans were at the center of things when it came to flexibility, innovation, and sustainability for long term. It is here, by way of contradiction, that the concept of Industry 5.0 originated, a definition of a new industrial future that rebalances by placing humans at its own center of technology.

Industry 4.0, best described as the age of cyber-physical systems, saw for the first time the integration of machines, sensors, and software via the Internet of Things (IoT). Through advanced analytics and artificial intelligence, factories were enabled to process large amounts of information simultaneously, even in real time. Production lines adjusted to demand shifts automatically, supply chains were optimized based on predictive modeling, and breakdowns were pre-programmed for maintenance. SMEs, for their part, saw hopes of competing even with large companies by embracing cheap digital technologies like cloud-based planning systems, machine vision software, and cheap IoT sensors. Industry 4.0 was all about automation, efficiency, and data-informed decisions.

Yet industry focus on pure efficiency occasionally proved to ignore broader worries. Industrialization for joblessness caused anxiety about employment losses, and a constant productivity push sometimes clashed with environmental protection. Besides, many firms, especially SMEs, also found it difficult to adapt to cultural and organizational shifts involved in embracing fully automatic systems. Industry 4.0 opened new possibilities—unprecedented—yet new challenges also loomed, which technology by itself couldn't fulfill.

This is where Industry 5.0 steps in. Instead of succeeding Industry 4.0, it builds upon and enhances it. Industry 5.0 places its focus on cooperative effort between humans and machines, not substitution. Instead of competitors to human labor, robots and machines are considered partners of theirs, means to enhance human creativity, decision-making, and solving puzzles. A collaborative robot (also known as a Cobot), for instance, is capable of working alongside a factory worker, relieving it of doing tedious lifting jobs while the human focus stays on precise assembly or quality enhancement. The process, besides being more productive, also enhances workplace protection and job satisfaction.

Another characteristic of Industry 5.0 is its emphasis on sustainability and social responsibility. As opposed to previous revolutions, which focused primarily on industrial development, Industry 5.0 acknowledges that manufacturing also needs to meet global challenges like climate change, resource constraints, and social fairness. For SMEs, it is particularly critical. Customers and governments are requiring sustainable behavior more and more, and Industry 5.0 offers the technological and organizational foundation for doing so. By employing artificial intelligence for energy optimization, renewable energy, or information twins for waste reduction, SMEs are able to decrease expenses and show environmental accountability at the same time.

Personalization is yet another industry 5.0 characteristic. With customers demanding products that meet their own requirements, SMEs are better placed to satisfy their needs. Their ability to be flexible and their small scale of operation enable them to be agile compared to large firms. SMEs are capable of delivering highly customized products thanks to collaborative software between AI and flexible production systems, which did not allow for small-scale production before and hence delivered a huge cost tag for it. For example, a miniature electronics manufacturer is capable of utilizing AI programs to create customer-order-specific circuit boards yet remaining efficient on the process line.

The shift from Industry 4.0 to Industry 5.0 of its kind is a great cultural and technical transformation. What machines could do, when linked and smart, we learned from 4.0. We are reminded of how irreplaceable are, yet, human creativity, intelligence, and accountability are, by 5.0. Both represent a harmonized model in which technology and humankind coexist, driving profitability as well as purpose.

For SMEs, it's a moment of both practical and strategic opportunity. Adopting Industry 4.0's tools of IoT, AI, and automation, and its industry's human-centric and sustainable ethos, SMEs have a chance to build a competitive advantage all their own. They are able to be as efficient as large corporations, yet still maintain their ability to bend, move quickly, and connect on a human level that their customers are demanding. That twofold benefit places Industry 5.0 not as some future aspiration, but as a concrete way forward for SMEs to prosper in the years to come.

1.4 Importance for SMEs

Small- and medium-sized businesses are the backbone of manufacturing globally. They dominate a majority of economies, representing over 90% of industrial establishments, and are major creators of employment, exporters, and innovators. Yet SMEs are usually hardest hit by modernization challenges. Large businesses usually have available funds, technical competence, and international linkages to adopt new technologies swiftly, but SMEs have to weigh each expenditure cautiously and usually suffer from a lack of funds. It is for this reason that smart manufacturing is of utmost value for small businesses. It is no longer a nice-to-have, but a survival issue.

One of the critical reasons why intelligent manufacturing is critical for SMEs is remaining competitive. Today's global manufacturing environment is highly competitive, where customers demand high quality, speed to market, and even customized products at reasonable prices. Large businesses are competing by employing sophisticated automation, digital supply chains, and planning systems based on artificial intelligence. If SMEs refuse to change by employing traditional approaches, they face being squeezed out of business by larger firms that are capable of delivering at a similar efficiency, thereby competing on cost and quality grounds. By embracing intelligent manufacturing technologies, SMEs achieve the capability to compete on a par, being as agile and efficient as larger firms.

Another is found in resilience. SMEs tend to work on thinner margins and smaller buffers than larger corporations, and hence are more exposed to disruptions. A pandemic like COVID-19, a geopolitical war, or even a drastic change in demand has proved how brittle supply chains are. Intelligent manufacturing software, like predictive analytics and digital twins, allows SMEs to see disruptions ahead, shift production timetables dynamically, and handle inventories better. For instance, a tiny auto component maker employing AI-based forecasting knows demand variations ahead of time and tweaks production capacity before inventory build-up or shortages happen. Such a measure of resilience can spell survival versus closing shop during a crisis.

Smart manufacturing is also critical for SMEs from the point of view of customer demand. It is no longer acceptable for customers to accept mass-produced, off-the-shelf products. Electronics, textiles, and consumer products are all demanding custom products and shorter cycles of products faster and faster. Standard manufacturing systems, based on fixed production lines, are not capable of delivering on demand by increasing production costs substantially. Smart manufacturing, based on flexibility and flexibility, means processes are easily reconfigured and custom products are delivered by SMEs efficiently. Offering custom products becomes a point of difference for a small business and a way for SMEs to build a niche for themselves amidst market clutter.

The sustainability role adds emphasis to smart production for SMEs as well. Regulators, governments, and customers are demanding greater carbon reduction, less waste, and green-business practices for companies. Large businesses usually have a sustainability division to manage the process, but SMEs have to reach a similar end through fewer people. Smart production is a viable solution here. Energy optimization through AI, waste reduction through data, and carbon-footprint tracking through numbers allow SMEs to achieve sustainability goals and achieve cost reductions at the same crossroads. It's one of the strongest smart-adoption cases based on a combination of environmental stewardship and cost efficiency.

It is also necessary to identify the cultural and organizational value of smart manufacturing for SMEs. By embracing digital technologies, small businesses are able to hire and keep young, technologically inclined individuals. Trained staff nowadays are interested in working where they will have contact with modern technologies, not traditional machines. Providing staff members with the chance to work with systems of artificial intelligence, cooperative robots, and digital platforms gives not only productivity growth but also a more inspired and interested staff. For SMEs, which often lack the ability to keep professionals, this is a key advantage.

Lastly, there is future-proofing. The rate of technological development does not appear to be diminishing anytime soon. What is advanced now will be commonplace tomorrow, and businesses that don't keep pace risk falling by the wayside. For SMEs, putting off investment in intelligent technologies might appear to be a short-term cost-cutting exercise, but on a longer timescale, it adds risk of being left behind to SMEs. By investing in getting their ducks in a row now, even with tiny incremental movements, SMEs can establish a foundation on which they are able to scale and adapt as technologies evolve.

Overall, smart manufacturing's value to SMEs cannot be understated. It helps SMEs be more competitive by bringing their feet into a parity game of economies of scale, creates resiliency against disruptions, facilitates customer-facing production, aids sustainability initiatives, enhances workforce buy-in, and guarantees long-term existence. Smart manufacturing for SMEs is not a growth strategy but the key to being a continued player in a growing digital, demanding world for SMEs. The adopters of it will be set to prosper, whereas its doubting Thomases may find themselves struggling to keep abreast of the future.

1.5 Glossary of Key Terms

As we talk about smart manufacturing, we encounter a lot of technical language that might intimidate, particularly for SMEs that cannot have their own IT or their own data science departments. The point is that it is really important to know those words, as they form the language of today's industry. Here, we are going to explain, in simple, useful language, the key concepts, explaining how their relevance applies to the SME reality.

Artificial Intelligence (AI) Artificial intelligence is the capability of machine or PC programs to undertake activities which require, under normal circumstances, human intelligence. Such activities involve pattern recognition, data learning, prediction, and even decision-making. When it comes to SMEs, AI is not fiction, it's a reality already ingrained in production data analysis, machine failure prediction, or job scheduling optimization of the most suitable type. Take, for instance, a metal fabrication SME, which may employ AI for predicting how it will require material for a period of some months ahead, thereby cutting wastage and expenditure.

Machine Learning (ML) Machine learning is a type of AI that enables systems to adapt on their own by learning from examples, rather than being directly programmed. Rather than being given a set of rules for each scenario, machine learning systems learn patterns and adjust their behavior accordingly. With SMEs, it could be a quality inspection system getting better and better as it processes thousands of images of products. It may also be a demand forecasting tool that tweaks as it discovers the seasonally related purchasing patterns of customers.

Internet of Things (IoT) Internet of Things refers to a group of physical objects, including machines, equipment, or sensors, which are linked to the internet and are capable of gathering and communicating information. Large factories link thousands of machines through IoT, but SMEs can gain advantages even by implementing it on a small scale. By fitting cheap IoT sensors on a few key machines, a business can gain information about energy consumption, vibration, or temperature, and identify issues early on, thereby averting pricey downtime.

Digital Twin A digital twin is a digital likeness of a physical item, process, or system. It is a virtual mirror of its real-world equivalent, in real time, allowing businesses to try things, model, and measure performance without disrupting live production. For instance, an electronics-producing SME might undertake a digital twin of its assembly line so it can model how a new design piece is going to impact output before actually changing things physically. It lessens trial and error expense and enhances choice-making.

Cloud Computing Cloud computing is the provision of computing services, that is, storage, software, and analysis, over the internet instead of local servers. SMEs don't have to purchase costly hardware by subscribing to cloud-based solutions, which scale for their requirements. Advanced technologies become economical even for tiny companies. A tiny SME, for example, may employ cloud-based enterprise resource planning (ERP) software for order tracking, inventory management, and production planning without massive upfront expenses.

Big Data Big data are huge and complex data sets that are not analyzable by traditional techniques. The data in manufacturing may originate from machines, sensors, suppliers, customers, and supply logistics networks. Even if the term is ominous, for SMEs it usually starts on a simple note of aggregating machine data and sales information onto a single dashboard. An SME that knows and responds to this data is better placed to know its inefficiencies, predict demand, and minimize waste.

Blockchain Blockchain is a record book of digital records that stores data securely and openly. Though blockchain is usually linked to virtual currencies, it is also forcefully applicable to supply chains. For an SME, blockchain is useful for guaranteeing materials that are sourced from trustworthy suppliers, following products end to end, and guaranteeing originality. A tiny food-processing SME, for instance, might employ blockchain to guarantee to customers where its raw materials originate and that they are of good quality.

Robotics and Cobots Robotics is the application of programmable machines for task accomplishment, usually those which are hazardous or routine. Cobots, or collaborative robots, are meant to operate alongside people safely. Unlike older industrial robots, which are typically caged, cobots are compact, adaptable, and for SMEs. They are easy to work and assist workers with tasks such as packing, welding, or assembly, resulting in less stress and increased productivity.

Cyber-Physical Systems (CPS) Cyber-physical systems integrate digital and physical processes, where computer algorithms are closely embedded in mechanical systems. That's how we have modern "smart factories," where equipment at a physical site is controlled and monitored by digital intelligence. For SMEs, even a low-level cyber-physical configuration, such as connecting machine sensors to an analytics dashboard, can change shop-floor visibility and efficiency dramatically.

Sustainability in Smart Manufacturing Sustainability is about activities to minimize environmental footprint and keep a process economically sound. With smart manufacturing, sustainability is facilitated by technologies to decrease energy, optimize, and minimize waste. Sustainability for SMEs is not merely about doing what it's told, but it also lowers expenses and gains reputation by going green for greener pockets and greener consciences of customers.

1.6 Practical Element: SME Readiness Snapshot Checklist

Comprehending smart manufacturing concepts is valuable, but for SMEs, it's necessary to convert it into real action. Until an enterprise embarks on its smart manufacturing journey, like going into Industry 4.0, or Industry 5.0, it first needs to honestly assess its current position today. It's a critical, yet usually avoided, step. Organizations are hasty, jumping to employing digital technologies even if they might not have a foundation to sustain them, resulting in money going down the drain and disappointment. A preparation test identifies SME's points of strength, realizes its holes, and determines its introductory steps for smart manufacturing.

The first focus is on being financially prepared. Intelligent manufacturing technologies are available across a spectrum of prices, ranging from affordable IoT sensors to sophisticated digital twin software systems. An SME cannot start for millions of dollars, but it cannot start without a sharp sense of how much money is available to spend and by when a cash payback is required. The owners should pose themselves questions like: Do we have money saved for tiny pilot initiatives? Are we also looking at spending money on tools that cannot give us an instant outcome but are going to change our competitiveness for the better by a long shot?

The other area is data quality and availability. Smart manufacturing runs on data. Without sensors on machines, without digitized production records, or inconsistent sales information, even the greatest AI software cannot produce useful information. One possible thought of an SME is as follows: Are we actually capturing data off our processes? If so, is it trustworthy and tabular? If not, what are some of those tiny steps we can start taking to measure what's most important—dwelt, energy, or defectivity?

Coming next is technology infrastructure. A few SMEs continue to maintain by hand record-keeping or ancient equipment that cannot readily be interfaced to current digital technologies. That does not preclude transformation, but it does preclude it starting at a revolutionary level—transforms must begin incrementally. A manager might ask himself/herself: Does our equipment permit sensors to be added? Does our business benefit from cloud computing, even local servers to store and manipulate data? Even small purchases, like adding low-cost sensors to existing equipment, are a possible place to begin.

A third critical element is people and skills. Technology by itself cannot achieve transformation, but it needs a workforce primed and motivated for new ways of doing things. For SMEs, a frequent need is for training and cultural shift. A helpful question to ask may be: Does our workforce value digital tools, or are they perceived as a threat? Have we committed to simple training to allow our workforce to work off new systems? Getting staff excited about things and trusting things are as important as deploying new hardware or software.

The leadership mindset is no less important. Most successful SME transformations start not with technology but with vision. Leaders need to challenge themselves: Are we serious about leading this change, even if challenges are looming? Are we crystal clear about why smart manufacturing is needed for our business? Without leadership commitment, even a cherished strategy will fail.

Lastly, SMEs should also think about external assistance and collaborations. SMEs, unlike large firms, which engineer, as it were, expertise for themselves, frequently have to depend on technology suppliers, consultants, or even university and research institution collaborations. With regard to this, managers should ask themselves: Are we comfortable having partners we trust who can lead us? Are we receptive to establishing cooperation with other SMEs or participating in digital transformation clusters to exchange ourselves for it?

By thoughtfully examining these aspects such as financial resources, data, infrastructure, workforce, leadership, and partnerships, an SME can build its own profile of readiness. There is no one score, nor a one-size-fits-all benchmark. Rather, it's about self-knowledge. A firm may find it has existing knowledgeable staff but it lacks suitable data gathering systems, or it has dedicated money to sensors but struggles with adoption by the workforce. Every discovery becomes a stepping stone along the transformation roadmap.

A readiness snapshot should not be a fixed-time activity. As technology advances, so should the self-reflection of a company. An SME, which evaluates itself through this process once a year, will be in a position to monitor how far it's gone, tweak its strategy, and gain confidence during its digital process.

Essentially, this preparation exercise turns things of abstraction into real direction. It affirms that SMEs are not required to set everything before they start. They just require insight into where they are and the bravery to start by taking a single step. Every superb change, anyway, is anchored by a moment of truthful self-reflection.

Chapter 2
Challenges Unique to SMEs

Introduction For SMEs, the advent of smart manufacturing is an inspiring and challenging experience altogether. SMEs form the backbone of an industrial economy and are characterized by their agility, innovation, and ties with the communities they serve. However, while adopting the latest available technologies of Industry 4.0 and Industry 5.0, SMEs are suddenly confronted with a dilemma. There is an opportunity to be more efficient, sustainable, and global, but the challenges appear to be expanding with each new technology.

The fact is that SMEs work with a much smaller budget than large corporations. The truth is that they cannot afford to shell out millions of dollars in terms of R&D, infrastructure, and employee training and development programs. The success of SMEs lies in making the right decisions: making effective expenditures, identifying those things that provide immediate value, and developing their digital acumen in a gradual manner.

The problems that SMEs experience are in no way a function of a lack of vision and skill. The smaller the company, the more agile and flexible it is likely to be than larger companies. The environment is the biggest issue that SMEs experience: difficulties in access to skilled people and funds, a dysfunctional supply chain, and, at times, an alien environment in terms of policies too. Moreover, the tools for the digital age are created keeping in mind large corporations, and SMEs find it difficult to fit in due to the costs and complexities involved.

The first step in the process of overcoming challenges is to be clear about the difficulties involved. This chapter will review the issues that serve as hindrances to the full benefits of Industry 5.0 and smart manufacturing for the SMMEs in the country. By analyzing costs, data, and the issues involved in integration and skill levels, this solution provides insight and a way forward. Its purpose is to encourage people to achieve a positive outcome.

A. A. Kadam, *AI-Powered Smart Manufacturing*, SpringerBriefs in Applied Sciences and Technology, https://doi.org/10.1007/978-3-032-21466-9_2

2.1 Cost Constraints and ROI Concerns

Cost is one of the major hindrances for small- and medium-scale manufacturing units to shift to smart manufacturing. Even though digital technology is less expensive than it used to be, it still requires initial outlay, which seems exorbitant for small businesses. Implementation of AI-based monitoring systems, IoT sensors, upgrading old systems, and cloud-based manufacturing software requires initial and secondary costs. For SMEs, where the profit margins are small, even the smallest of investments require measurable benefits.

Many times, the fear is that costs are higher than they actually are. Small businesses will feel that adopting new technology means investing huge amounts of money in new equipment and waiting for a very long time to recover their expenses. But smart manufacturing technology does not always have to go for the overhaul of the existing systems that the business is using. It could simply begin with small steps. For instance, the business could start by installing a number of sensors on the most important machines, but not invest huge amounts of money and time into the automation of every machine in the business. Even that will provide important data on how the machines operate, the maintenance they require, and the amount of energy they consume.

Return on Investment (ROI) remains a huge concern since small- and medium-scale businesses operate within strict financial parameters. Each dollar invested needs to bring lasting benefits, and these should materialize within months, not years. However, savvy manufacturing technology may bring benefits that compound over time, not instant, massive savings. This results in a huge disconnect between the financial benefits that need to accrue and the benefits that the technology brings. This suggests that business owners need to shift their paradigm regarding the adoption of technology, regarding it not as an expense but as an investment.

For instance, take the case of predictive maintenance. It may appear costly when one installs sensors and analysis mechanisms that monitor the functionality of the equipment. However, the predictive maintenance technique helps organizations avoid downtimes, which can save them thousands of dollars that they could have incurred. It is the same with the use of AI-based scheduling software, which appears unnecessary but reduces the time an organization spends idle and brings higher profits.

SMEs can also attempt innovative approaches for project financing, as well as collaborate with other organizations for reduced initial expenses. Additionally, governments, along with various sectors, provide grants, subsidies, or even tax breaks for assistance with digital transformation. Lastly, technology firms also provide subscription-based services—pay-as-you-go—which lessen significant capital outlays. Cloud computing services allow SMEs access to sophisticated software without purchasing it, paying for the time used or length of use.

Cost concerns are still a worry that is alleviated by not having specific numbers when it comes to money. ROI is easy to measure for regular machines. However, digital technology provides benefits that are not that easy to measure, such as better decision-making, avoiding waste, happier employees, and happier customers. Many small- and medium-scale businesses will not take the leap of faith because of this.

To forge ahead, it is essential that small- and medium-scale businesses adopt a phased approach. By undertaking a small project, analyzing the results, and growing incrementally on the basis of the data that is generated, businesses will eventually reach the stage of adopting transformative technology. By deriving actual benefits, business leaders will gain the confidence of their own organizations for undertaking transformative purchases.

Ultimately, the cost of technology is not simply about expense. It's about how businesses think about technology. If small and medium businesses break down the transformation journey into smaller steps, they will not feel overwhelmed by the cost. Smart manufacturing does not have to happen correctly on the first attempt. It only requires consistent and informed progress. Every sensor that goes in, every data point that is collected, and every small increment bring the business closer to resilience, efficiency, and readiness.

2.2 Lack of Skilled Workforce

Even if the factories are run by technology, people are the ones who keep the factories running. No matter how high the level of sophistication reaches, the systems are useful only when the people who use them are capable, competent, and adaptable. For small businesses, the biggest problem is that people are required to run the technological systems.

The problem that small- and medium-sized businesses are facing worldwide is that there is a huge skills gap that exists in these enterprises. The skills that are required are different from the ones that the workforce has. In the past, the way in which manufacturing worked was that people acquired skills from hands-on experiences. This is because a lot of machines used in factories were mechanical in nature, and changes in the nature of manufacturing used to take a long time. The emergence of smart manufacturing has changed the scene, as now the workforce is required to analyze data, work with machine learning, make use of automation, and also interact with robots.

The problem for small to medium businesses is that they lack easy accessibility to training programs. Large businesses can afford to set up a training center within the company and fund to send people for professional training, but small businesses lack the means to train their people. Training is a cost in terms of money as well as time, which small businesses lack in abundance. By way of illustration, a machine operator is working overtime; he cannot spend a week learning about the automation of industries. The small factory owner might find it difficult to send people for training when the return on investment is not apparent.

The age difference between experienced and young employees intensifies the problem. Smaller firms, for instance, have experienced technicians who are familiar with the equipment but are not comfortable with digital solutions. The young workforce, on the other hand, is competent with technology but lacks familiarity with the industry. The best way to blend the two is with a mixture of experience and innovation—something that cross-generation mentoring would accomplish, with the senior workforce teaching the process, while the youth assist in implementing technological solutions.

The other side of the skills gap is that of digital literacy. Even rudimentary skills, such as the use of cloud dashboards, data visualization, or collaboration tools, are commonly in short supply for small- and medium-sized enterprise (SME) employees. In a smart factory, sensors are constantly churning out bits of information, which are then to be read, identified, and acted on by the workforce. In the absence of such skills, smart manufacturing doesn't really live up to the promise that it holds.

The lack of expert labor hampers the effectiveness of the adoption of modern technologies such as AI or robotics within SMEs. For instance, an SME might acquire a collaborative robot (cobot) that is meant to undertake repetitive tasks such as in the assembly line. However, a lack of people with the skills to operate, maintain, or program such a robot means that the robot is underutilized. Similarly, a factory might install sensors meant to support IoT for the acquisition of certain types of data but lacks people who are capable of interpreting such data.

Solution requires more than short-term training; it requires a cultural shift. For small- and medium-sized enterprises (SMEs), learning needs to become a part of an ongoing process, not a single event, which can be incorporated into their daily activity. Solutions such as short video tutorials, learning sessions led by colleagues, workshops, or industry support, even from vendors, would help people learn on a constant basis without hampering production. The government plays a significant role here. In most nations, regional technology organizations provide affordable, even heavily subsidized, training specifically developed for small fabricators. Such initiatives ensure that small businesses acquire knowledge that would be financially unaffordable for them.

Equally significant is reskilling and redeployment. This is because automation alters the way in which individuals do things, but it doesn't eliminate employment; it merely changes it. This means that tasks that were formerly dependent on human accuracy are now performed by computers, but new tasks arise that encompass watching computer systems, interpreting computer-generated results, and optimizing processes. This is particularly significant for small- to medium-scale businesses because, rather than being afraid of automation, owners need to find a way to empower existing employee skills to keep pace with automation. This means a technician who formerly adjusted computer settings is now capable of handling computerized control systems.

Lastly, recruiting fresh talent for small to medium manufacturing businesses is highly essential. Young people perceive small manufacturing businesses as outdated compared to tech startups and service businesses. For such businesses to appear more appealing, they need to modernize themselves. If potential recruits see a smart dashboard, collaborative robot, or use of digital technologies in the factory, it can make a huge impact on how they perceive the business. This is because, in addition to enhancing productivity, smart manufacturing helps individuals with necessary skills to join the business.

In sum, a lack of enough qualified personnel is more than a matter of recruitment. It holds back competition, innovation, and success. For small to medium businesses to succeed in Industry 4.0/5.0, skills development has to become part of their strategy, similar to investing in equipment. Technology, without people who can operate it, is a dead end. But with the requisite skills, a small business can be at its best. The future of intelligent manufacturing belongs to the most adaptable businesses, which begins with a skilled and confident workforce.

2.3 Data Availability and Quality Issues

In modern smart factories, data is considered money. It drives AI and automation and informs all decisions on the factory floor. The problem is that many small- to medium-sized businesses do not make use of their data. Big businesses spend a lot on gathering, storing, and analyzing their data, but small businesses have data that is fragmented, incomplete, or even non-existent. The problem of a lack of credible data is a significant factor that hinders success for small businesses in undergoing digital transformation.

The problem is that most small and medium enterprises are not set up to be data-driven. The systems that they have are the result of a gradual process, with a lot of things fixed on whiteboards, spreadsheets, and on paper. The production plans are written on whiteboards, quality is documented in notebooks, and inventory is counted by hand at the end of every day. This is how a traditional factory is set up. This is not how a smart factory is supposed to operate. Here, automation has to interact with people, software, and machines on a constant basis, which is not possible when there is no structure in the data.

Even when small businesses are handling the data, a problem with the quality of the data might occur. The entries may be inconsistent, there might be incomplete information, and even errors generated when people type by hand might compromise analysis results. For instance, when machine downtime is not recorded in a similar manner every time, when the production numbers are only an estimate, the analysis might provide misleading information regarding efficiency. In smart manufacturing, bad data is worse than no data because people might become overconfident.

The problem is that, in small- and medium-sized enterprises (SMEs), the data is fragmented. The information is in different systems, or even in people's minds. For instance, machine data is in one spreadsheet, quality data in another, and maintenance data in a third. Having such fragmented pockets of data prevents us from having the full picture. We cannot know whether certain process conditions are causing problems with the results, because we are not connecting the machine performance with quality results.

The environment of data handling is also an important consideration. In small to medium businesses, the people involved may not clearly recognize that there are benefits to following a consistent approach to gathering data. Efforts such as tracking each event, providing process details, or qualifying dashboards with updated information might be perceived as unnecessary tasks with little, if any, return on investment. It is understandable when making data-informed decisions is still non-standard practice for a business. It is essential to communicate how improved outcomes are obtained with the use of data, such as when precise downtime reports enable faster maintenance, for instance, or scrap rate reporting informs process optimizations.

Data integration is a further layer of complexity that arises. In a small- to medium-sized enterprise (SME), a variety of old and new machines may be in use, with varying methods of intercommunication. It is common that the old machines, which are still working efficiently even after a couple of decades, operate remotely via non-electronic interfaces, which means that there is no use of modern computerized interfaces to fetch the required data. Although the implementation of sensors on such machines, as described later, can prove useful, it still requires significant knowledge and investment. The best thing is to take a small, pilot, high-value approach to get the desired network started.

Cybersecurity concerns are also limiting the amount of data that is collected and shared. Smaller businesses have been extremely cautious when it comes to connecting devices to the internet. This is because they are concerned about breaches of data confidentiality. Then again, the solution to this problem is not keeping the devices offline. There are now safe platforms available that can be used by SMEs, thanks to cloud service providers and IoT suppliers. This helps keep the data usable while keeping it safe as well.

Despite the presence of such challenges, it is essential to keep in mind that data management is not a matter of either/or. The transformation to a data-driven enterprise can begin with tiny steps. An industry, for instance, can begin by digitizing the records of production with basic software applications or by the use of low-cost sensors that monitor the operation hours of machines. Over a short while, such small chunks of data increase and identify certain trends such as bottlenecks, energy inefficiencies, or quality issues. Then, when a robust foundation of high-quality data has been created, more sophisticated technologies such as machine learning may be introduced.

For small to medium businesses, data transformation is more than a technological challenge. Its implications are cultural, although it requires leadership, employee participation, and a mindset of openness. Data is a resource that facilitates improvement, not merely a by-product of a business. If small businesses believe in their data and keep it complete, precise, and on time, then, in effect, they are in complete control of the business.

In conclusion, the availability and quality of data are not merely a matter of technology but a matter of mindset, discipline, and a long-term vision. The smart factory is not realized overnight but is cultivated gradually, one measurement at a time. Even a little activity toward the collection, cleaning, and connecting of data in small to medium businesses can bring huge changes. "You cannot improve what you do not measure." That is exactly what happens in smart manufacturing; people who measure better and manage their data better are bound to produce the success stories of the future.

2.4 Technology Integration with Legacy Systems

For small- to medium-sized businesses, a shift to smart manufacturing is not necessarily an implementation from scratch. It is an implementation that begins in factories with a lot of old equipment that has been working perfectly for a long time, even prior to the existence of the "digital twin" or "Industry 4.0" concept. Such equipment exhibits a great amount of past investment and past knowledge, and, importantly, performs the mechanical part of the job very well. The conflict arises when such equipment has to work together with the latest digital technologies.

The problem that arises is how to make the old systems, those that are not digitized, work side by side with the new ones, which is, in my opinion, one of the biggest challenges that SMEs have when embarking on smart manufacturing. Large corporations can simply remove entire production systems that are not digitized, replacing them with modern, digitized ones, but in most cases, SMEs cannot afford to remove equipment that is still working, even though it is not digitized. Hence, the term "brownfield," in contrast to "greenfield," is used when designing entirely digitized factories.

The problem is that most of the older machines were not designed to communicate digitally. The control systems on most machines are closed, the interfaces are outdated, and the data is locked into analog signals or proprietary messages. Take, for instance, a CNC lathe that an SME might own that is 20 years old. The lathe might produce very high-quality products, but it would not be able to provide real-time information related to the spindle speed, tool life, or production cycles. The machine would be working properly, but it would not be visible to the different digital systems used in modern manufacturing.

The bright side is that contemporary technology has evolved to find a solution to this problem. One technique that has proven to be highly useful is retrofitting, which is the addition of sensors, controls, or components of an interface to existing equipment to enable the interpretation of how that equipment is working. Sensors monitor such things as vibrations, temperatures, current flow, or cycle periods and convert analog signals into digital signals that computer software can interpret. This helps small to medium businesses because retrofit kits are becoming less expensive, making it easy to begin small.

The other solution is to apply devices that are small, intelligent, and installed between legacy systems and modern networks. The devices are capable of extracting data from legacy systems, interpreting it, and then sending useful results to cloud databases for storage purposes. The use of edge computing is essential for small to medium businesses because it provides real-time data without straining the network. For instance, a small business can employ an edge gateway that extracts signals from different machines, which are then normalized before sending the reports to an overall dashboard.

Integration requires the issue of software compatibility to be resolved. Smaller businesses, for instance, are known to use outdated enterprise software or even spreadsheets that were not originally interfaced with IoT platforms and AI software. Here, middleware is useful. It is software that facilitates the interfacing of different systems. For instance, it is possible to make schedules from an old enterprise resource planning software communicate with sensor data from a factory floor, resulting in improved coordination. Such coordination is a result of a smooth workflow that is facilitated by the middleware software.

It is also worth considering how a plan for gradual investment might work. This is particularly relevant because small and medium businesses are unable to implement solutions in all domains simultaneously. They can, therefore, prioritize solutions, which might include identifying machines that are most essential for the business, whether because they are most used in the production process or whether they are more prone to malfunctioning. It is, therefore, easy to see how small businesses might quickly derive a number of benefits from going digital.

Security needs to be considered in every integration strategy. The danger of upgrading existing machines to connect to the internet is that existing machines would never have considered a need for such a mechanism in the past. Network segmentation, segregating the vital equipment used in production from the internet, is a strategy that small to medium enterprises need to apply to ensure safety. Using a safe gateway is also essential in regulating the flow of data.

There are numerous instances of small to medium businesses leveraging legacy integration. For instance, a small car parts manufacturer in Germany installed IoT sensors on top of existing 15-year-old milling machines to track spindle vibration and temperatures. In a matter of a few months, they reduced unscheduled shutdowns by 30% and identified that two machines were consuming more power compared to other machines with varying loads. This assisted them in making proper maintenance plans, which yielded a return on investment for sensors within 6 months. Another instance is that of a medium-scale textile manufacturer in India, who implemented a cloud-based dashboard that synthesized data from both modern dyeing machines and the other, older ones.

The essential message for small to medium businesses is that legacy systems do not hinder progress but provide an opportunity. In trying to digitize what is already there, rather than replacing it, small to medium businesses can keep their cash, prolong the life of proven machines, and still leverage the power of smart manufacturing intelligence. This means that smart manufacturing is not eliminating what came before but connecting what came before with what is going on now. Every legacy machine that is connected to the smart network is a valuable source that helps a business be clearer, more efficient, and nimbler.

In the long run, how effectively small and medium enterprises can interconnect the existing systems will indicate how effectively they can connect the Industry 5.0 network. Such factories that have the ability to interconnect various generations of technology, combining the best of reliable systems with the best of knowledge, can have the best of both worlds. This way, they can keep their existing systems alive and make them a source of innovation. Thus, integrating technologies becomes a brilliant business strategy.

2.5 Regulatory Pressures

In modern industry, the regulatory environment used to play a behind-the-scenes role, but now it's influencing how businesses operate. This is extremely significant in terms of small to medium businesses. Regulations impact nearly every aspect of the industrial sector, including product safety, working conditions, the environment, and cyber protection. Smarter industries mean that regulations are becoming more complicated, with even more obligations for small businesses that are tight on time, people, and capital.

Traditionally, regulations concerned with physical outputs involved ensuring that the product is of high quality, ensuring that there are no accidents in the workplace, and ensuring that trade practices remain fair. The impact of technological evolution in the industry has significantly increased the number of regulations that small to medium businesses need to adhere to. For instance, data protection regulations, such as the GDPR, are now applicable to manufacturing businesses that exchange operation data. Additionally, environmental regulations, such as ISO 14001, are now more stringent because of government advocacy for environmentally friendly businesses. Even though regulations are essential within an industry for purposes such as safety, sustainability, and fair trade practices, a small business may find them burdensome.

Often, small- to medium-sized businesses (SMEs) are unsure about what regulations they, as businesses, are supposed to be following. The regulations depend on the sector, zone, and markets involved. For instance, a small manufacturer of auto parts, which is involved in the export of auto parts to Europe, has to follow the regulations of REACH (Registration, Evaluation, Authorization, and Restriction of Chemicals), as well as RoHS (Restriction of Hazardous Materials). For an SME that is a manufacturer of food, HACCP (Hazard Analysis and Critical Control Point), which is a food safety regulation, has to be followed. This includes traceability, which takes a significant amount of time.

The application of digital technology can make it more difficult to comply, although it is very useful in other aspects. When small- and medium-sized businesses (SMEs) begin to use IoT sensors or AI for analysis, they produce a significant amount of data concerning their businesses, as well as sometimes concerning people. This might involve machine logs, production rate, who accessed what, and business with suppliers. In order to manage this type of data properly, they are obliged to respect the regulations concerning data protection and cybersecurity. Indeed, failure to protect the data appropriately might result in non-compliance, penalties, and lack of consumer trust. Most SMEs lack the necessary know-how to master such complicated regulations.

Sustainable development is a significant area, with regulations becoming stricter in the current environment. National governments are imposing stricter emissions regulations on businesses, which are encouraging them to calculate and state the carbon footprint. Smaller businesses, which use outdated equipment, might find it difficult to cope with such requirements. Installing energy-conserving systems, waste management, and pollution control can be highly expensive. It is not an alternative to opt for a non-compliance solution anymore. Customers, particularly MNCs, are encouraging suppliers to prove they are environment-friendly businesses as part of attaining sustainability worldwide. Smaller businesses, even locally operated, are forced to adhere to international regulations because of such challenges in the supply chain.

In these instances, smart manufacturing technology can be of assistance. Digital solutions enable SMEs to track compliance metrics remotely. For instance, energy use or emissions can be tracked real time, which means that issues are identified before they become non-compliances. Compliant reports can be generated by leveraging analytics, which means that the burden on the administration staff is reduced. Cloud solutions can enable the storage of documents, ensuring that audit reports are up-to-date. In these instances, non-compliance can be a means for improvement.

Compliance is even more than ensuring proper equipment is in place; there's a need to develop a compliance mindset. The fact is that, for far too many small businesses, regulatory issues are dealt with only when the time for audits or inspections arises. This is a recipe for stress, inefficiency, and even failure. On the other hand, when a small business embeds regulatory thinking into quality management, product, and supply management, complying becomes a part of regular business.

Industry associations, as well as government organizations, assist small- to medium-sized enterprises (SMEs) during the transition process. Today, assistance with regulations, resources on the internet, and reduced costs for certification are available in many areas. For instance, a small electronics manufacturer in a country in Southeast Asia used a traceability initiative made available by a government agency to assist with European importing regulations, which increased the number of customers from abroad. On a similar note, environment grants in a number of European nations assist with the cost of energy conservation systems/renewable energy equipment for SMEs.

The compliance process as a whole has to be owned by the SMEs. The best way to implement this is to identify a compliance champion, who is an employee of the organization, responsible for identifying changes in regulations. The employee doesn't necessarily need to be a lawyer but should possess strong organizational skills. Following the shift toward becoming a more digital business, this can develop into a leadership role on data governance, sustainability, etc.

Transparency is also important in the SME sector. In the modern age of computers and technology, transparency is what customers, as well as the regulatory bodies, are expecting. It is a great way to increase trust, which helps to increase your reputation when the transparency is in respect to compliance, sustainability, ethics, etc. Companies that tend to maintain high standards before the onset of requirements find it easy to enter into partnerships with larger businesses.

The regulatory challenges that small- to medium-sized businesses (SMEs) face are a serious hurdle on the journey toward smart manufacturing. But regulations can also drive changes toward modernization. By overlaying regulations with smart systems, remaining proactive, and leveraging government assistance, regulatory challenges can become a competitive advantage for a small- to medium-sized business. The businesses that succeed under Industry 5.0 are not ones that disregard regulations, but ones that leverage regulations as a foundation for innovation, safety, and sustainability. It's essential for small- to medium-sized businesses to remain proactive, not just on regulations for today, but on preparing for tomorrow's opportunities.

2.6 Practical Checklist: Assessing SME Readiness for Smart Manufacturing

Prior to embarking on a path toward smart manufacturing, it is essential that each small to medium enterprise take a step back to assess exactly where they are now. It is simple to covet the latest technologies such as artificial intelligence, automation, robotics, or the concept of a digital twin, but it is essential that a clear understanding of the current state is available, lest there be a misalignment of objectives with underwhelming outcomes. Readiness is not a matter of passing or failing but a means of self-discovery.

The very first thing that a small to medium business should consider is asking, "Why do we want to transform?" The reason for transforming affects the plan. Either improved daily working, improved product quality, reduced waste, expanding markets, or maybe a combination of all these is desired. There is no best answer to that. The most essential thing is that it's clearly stated. For instance, a factory that wants to reduce idle time requires different technologies and skills from a factory that wants to produce customized goods.

Having identified the reason for change, attention now moves to leadership and commitment. Any change requires a leader who is committed to that change. In small and medium businesses, the leadership style is very personalized because the owner, who is the managing director, stamps his/her authority on the entire outfit. This leader has to approve changes in budgets, apart from championing the cause for change. This requires articulation of the vision, ensuring that employees recognize that technology is a helper, not a threat, within the workplace. This leader also has to demonstrate a readiness to learn, hence encouraging his/her workforce to do the same. The question arises, therefore, of whether, as leaders, we are ready to change before asking the workforce to go along with us.

The third factor is a matter of organizational culture. In smart manufacturing, different sectors such as production, maintenance, quality, procurement, and IT have to come together, even if they are broken up into different departments for several years. Smoother flow of data is essential, and shared decision-making is required. Small- and medium-sized enterprises need to consider whether their organizational culture supports collaboration across different functions, whether people feel empowered to make suggestions for improvement, whether people feel encouraged to speak up, etc.

An even more significant consideration, though, is a competent workforce. This chapter has already stated that a skill gap is a massive hindrance for SMEs. In this readiness, it is necessary that the workforce is not a team of data scientists but people who are generally comfortable with technology and driven to learn. This, of course, can be initiated by identifying digital leaders within the current workforce, people who are inherently interested, enthusiastic about trying new things, and admired by their colleagues. The most obvious question that arises is this: Are our workforce members equipped with the skills necessary to operate and maintain these technologies when introduced? If not, are we prepared to develop them?

Data readiness is the next step and the cornerstone of smart manufacturing. Most small- to medium-sized enterprises underestimate the importance of data integrity. Prior to investing in analytics solutions or AI, ensure that your operational data is being gathered, archived, and processed in a consistent manner. This requires solid tooling, but even more, sound practices. The best way to proceed is to digitize your production records, normalize your data formats, and centralize information now contained in spreadsheets or on paper. Leaders need to assess whether they can trust their own data, whether it's reliable, timely, and complete enough to support better-informed decisions.

Technology infrastructure is another vital foundation. In intelligent manufacturing, machines, sensors, and software need to communicate with each other. It is essential for small- to medium-sized businesses (SMEs) to gauge whether their existing equipment is capable of integrating with one another. It is not essential for businesses to upgrade when existing machines are outdated, but such machines should be capable of collecting data in the future. The IT systems used within the business, such as EP software, need to be robust, secure, and scalable. The most fundamental question here is: Is our equipment capable of connecting with each other?

Being financially ready is essential for how quickly and how far a small to medium business wants to develop. In intelligent factory investment, a financially prepared small to medium business means it is not only aware of the cost of the technology, but also of how long it takes to pay back, besides other benefits that are indirectly derived. It is essential for businessmen to examine their finances very carefully before exploring all means of financial assistance available in the form of government assistance, suppliers, or sectors that assist in digitization.

Readiness assessment is never complete without cybersecurity and risk management. The more factories are connected, the more cyber threats rise. Even a single incident can bring production to a standstill, erode trust, and result in financial loss. SMEs should implement the following fundamental cybersecurity practices: use of robust passwords, restricted system access, frequent software updates, and employee awareness. Managers need to consider the following questions: Would we be able to quickly respond in case our systems are hacked today, and safeguard our critical information?

It is necessary that SMEs look at their environment. The smart factory benefits from collaboration. It is a fact that suppliers, customers, and tech vendors are helping to develop value chains. For small businesses, membership in industry clusters, innovation hubs, or universities provides access to knowledge that is not in their possession. The challenge is whether we are leveraging our environment to accelerate our transformation journey or whether we are doing everything on our own. No small business should embark on Industry 4.0 on its own.

If you take a fair look at these sectors like leadership, culture, skills, data, technology, finances, cybersecurity, and ecosystem, the overall effect is what gives you a complete snapshot of your readiness level. There are some SMEs that might have strengths in certain sectors but may lack in other areas. This is perfectly fine. The aim of this self-analysis exercise is not scoring but charting a plan. An SME with sound technological infrastructure but lacking skills in people should plan on developing skills before resorting to automation. An SME with a passionate team but weak data infrastructure should plan on digitizing its archives.

Ready now is not a fixed position, but a position that evolves with the passage of time. The small or medium-sized business that is ready now may look back at such a checklist a year later and discover new strengths or challenges. The most critical aspect, therefore, is keeping with the times. Being a smart, connected, and resilient business is a continuous journey, not a destination for overnight success.

In a nutshell, checking readiness is a leadership activity that remains humble. It is a process involving looking within and then looking outward. It is a process that converts doubts into directions and ambition into plans. The SMEs that are serious about this process of self-reflection are bound to find the path to smart manufacturing with fewer risks and more assurance. It is basically the beginning of becoming a smart manufacturer that is, obviously, really smart.

Chapter 3
Myths and Real Opportunities

Introduction Every major shift in the industry brings with it a degree of excitement as well as a degree of concern. The use of electricity in factories during the Second Industrial Revolution saw a fear of job loss to machines. The use of computers in factories during the Third Industrial Revolution questioned whether humans would still be relevant in factories. The Fourth Industrial Revolution, with the use of smart factories, artificial intelligence, and machines, is following suit. The concepts of efficiency, precision, and sustainability are clouded with skepticism, misinformation, and fear, particularly in small businesses.

In small- to medium-sized businesses, myths can mean a lack of progress. They can mean a halt to innovation, a slowing of investment, and a fear of complexity. The fact is, smart manufacturing is not only for large businesses. It is a versatile strategy that can assist even a small factory with a smart plan that takes real needs into consideration. The cost of entry is far lower than most people imagine. Even small businesses can implement a series of smart solutions by leveraging affordable sensors, cloud computing, and modular AI solutions.

This chapter explores myths that exist in smart manufacturing and Industry 4.0, distinguishing what is real from what is not. This provides a way for SMEs to base decisions on what is real, what is not, and what is hidden behind myths.

3.1 Myth: Smart Manufacturing Is Only for Big Corporations

Repeatedly, one hears that smart manufacturing is only for giant international corporations. This has prevented small- to medium-sized enterprises from even exploring the realm of digital transformation, setting a psychological barrier before any cost-related barriers are even introduced. For small business owners, the phrase

A. A. Kadam, *AI-Powered Smart Manufacturing*, SpringerBriefs in Applied Sciences and Technology, https://doi.org/10.1007/978-3-032-21466-9_3

"smart manufacturing" brings to mind giant automated factories, highly trained engineers, complicated computer systems, and multi-million-dollar risks, which are far removed from a small factory or workshop setting.

The reason for that is this vision did not come from a void. In the beginning of Industry 4.0, intelligent manufacturing cost a lot of money and had a high implementation barrier. Only large corporations could afford that, which is why they were the front-runners in implementing intelligent manufacturing technologies such as state-of-the-art robotics, special software, private data centers, and data scientists. Because of that, the articles in the media that reported on success stories centered on large car manufacturers, aerospace, and large electronics corporations. But that is no longer true with the current state of technology.

The past decade has been a transformation era for smart manufacturing. The technologies that were once centralized, private, and costly are now modular, cloud-hosted, and accessible. Artificial intelligence is no longer confined to research institutions; today, it is a software platform that can be accessed with a monthly fee. The IoT sensors that once needed bespoke installation can now be easily installed on machines in minutes. The dashboards that once required a massive IT infrastructure are now available on a smartphone screen.

The biggest mistake is believing that the size of scale is the same as intelligence. Smarter manufacturing is not dependent on factory size. It's dependent on how intelligent the factory is with the way it makes decisions. For instance, a small factory with ten machines that is aware of all the factory's activity, such as downtimes, energy, and quality in real time, is smarter than a large factory that uses outdated reports.

For small and medium enterprises, the gap is large. For instance, a small manufacturer of metal parts doesn't require a fully automated factory to derive the benefits of smart manufacturing. Installation of vibration sensors on merely two critical machines is sufficient to reduce machine failures significantly. For a textile business, an SME doesn't require the use of AI throughout the factory; a single computer vision application inspecting textile defects can bring significant improvements to fabric quality. An SME in food processing doesn't require a massive blockchain infrastructure; a simple digital traceability solution can significantly increase adherence and customer satisfaction.

The reason why smart manufacturing is considered a good fit for small- and medium-sized enterprises (SMEs) is that they can implement it incrementally. Large corporations tend to have massive projects that take a lot of resources to execute. This is because for large corporations, a smart factory implementation takes a long time, hence requiring more coordination and significant capital investment. In small businesses, the implementation cycles are short because they operate on a small scale, which makes it easy to see the results, possibly even faster than large corporations in some cases.

This is also supported by research. In fact, many articles from journals such as the Journal of Manufacturing Systems, and even the journal *Procedia CIRP*, indicate that small- and medium-sized enterprises that use predictive maintenance, artificial intelligence for quality inspections, or digital scheduling systems tend to pay back the cost in a matter of months, not years. Large businesses, on the other hand, take longer because of the complexities involved in such businesses. Clearly, smart manufacturing is more economical for small businesses than large ones.

The other reason that such a common myth is still so widespread is that it is a belief that experts are necessary when it comes to smart manufacturing. Although high-level solutions are useful when it comes to working with experts such as data scientists and automation engineers, smart manufacturing solutions are becoming more and more user-friendly, with solutions such as visual dashboards, no-code platforms for artificial intelligence, and support offered by vendors that make it easy for small- and medium-sized businesses to take advantage of advanced analytics even when they don't know how to code.

There's a strategic reason to consider here too. The belief is that big corporations will always have the edge over small ones when it comes to technology, making investment unnecessary. The advantage that small businesses have is that they are agile. Large businesses need to optimize their process on a large scale, which includes plants, geos, or product categories. The small business can develop solutions specifically for itself. An electronics manufacturer can develop AI models specifically for a product line, while a medium enterprise involving a machine shop can optimize solutions for a set of customers that are limited in number.

The most significant aspect of smart manufacturing is that small- and medium-scale businesses can now compete on an equal footing. In the old-fashioned way of manufacturing, large businesses had an advantage because of their size. They used to save on cost, have stronger negotiation power, and have access to the international market. The race is now going to be based on how smart, adaptable, and fast a business is. It means that a small business that is smarter and adaptable can win against a larger business that is slow and rigid.

That smart manufacturing is for big businesses only is a practice that came from outdated mindsets concerning cost, complexity, and capability. The fact is, though, that smart manufacturing is not a rarity; rather, it is a resource that everyone can leverage. The reason why some businesses are succeeding with it while others are only hesitating has, therefore, nothing to do with size, but with mindset. Small-to-medium organizations with a mindset that believes that the journey to digital transformation is a never-ending journey find that smart manufacturing integrates into their daily practices seamlessly.

The truth is that the real answer is not whether small to medium enterprises can afford the use of smart manufacturing. The real answer is whether they can afford to disregard the use of smart manufacturing. The reason is that globalization is becoming more competitive, and the quality requirements of customers are increasing, as are the regulations pertaining to sustainability. The need is for intelligence, which is no longer a luxury but a need. The application of smart manufacturing is not a matter of becoming larger; it is a matter of becoming better, in which small to medium enterprises can find themselves at a distinct advantage.

3.2 Myth: Automation and AI Will Replace People

Among the most common misconceptions that have instilled fear within the manufacturing sector, the replacement of human labor by automation and artificial intelligence has been the most feared. In the case of small and medium enterprises, such a belief is even more common because, apart from being employees, the workforce has knowledge that includes craftsmanship, customer connections, and corporate knowledge. The apprehension caused by the potential replacement by technology is a barrier to digitization.

This fear has historical background. In previous automation, especially Industry 3.0, some human tasks were automated, mostly involving tasks that are repetitive or very tough. Thus, automation is linked with job loss in people's perceptions of industry, then came artificial intelligence, which intensified this fear. The media portrayed AI as a system that would substitute human thinking, not assist human thinking, thus solidifying that "smart factories" would be factories without people.

However, such a perspective overlooks what is capable with AI, as well as how work is actually practiced in small- to medium-sized businesses.

Artificial intelligence is very good with regard to dealing with a large amount of information, identifying hidden patterns, executing repetitive, precise tasks, and being dependable. On the other hand, artificial intelligence lacks context, moral sense, creativity, empathy, and the sense to tackle uncertainties on its own. The manufacturing sector, especially small- to medium-scale businesses, always has a constant requirement for such skills in human beings. Customized orders, changes at short notice, client-specific requirements, and on-the-spot solutions are required on a daily basis in small to medium factories. This is exactly what cannot be replaced.

In fact, smart manufacturing doesn't eliminate people from a job. It alters their job. Machines do repetitive, tiresome, and error-prone work, while people supervise, decide, and optimize. For instance, rather than examining hundreds of similar things manually every shift, a person might be responsible for supervising the AI vision system, researching problems identified by the vision system, and collaborating with engineers to identify and rectify problems at their root. The work is no longer tiresome but more analytical, more useful, and more dignified.

This is evident in small to medium businesses because of their flat structure, which makes them very agile, unlike larger corporations. In small businesses, it is common to see the same employee operate a machine, address quality issues, and interact with customers. The emergence of AI systems means that these individuals are not replaced but made better. The AI system's dashboard would be used to monitor the performance of machines, which would help the operator make decisions faster and better.

Industry 5.0 focuses on human–AI collaboration as a way to shape the future of the manufacturing sector. Unlike Industry 4.0, which concentrated on automation, Industry 5.0 indicates that a sustainable development of industry is only possible with the use of human creativity, as well as with respect for the well-being and social responsibility of people. In this regard, AI is not making decisions on its own. It is a copilot here.

Collaborative robots, or cobots, make this concept evident. Cobots are intended to assist with tasks together with humans, not behind a safety enclosure. They assist with tasks such as lifting, positioning, repetitive tasks, or precise tasks, which reduces the tediousness of a physical job and the likelihood of injury. In small- to medium-sized businesses, which lack the means to emphasize ergonomic design with safety for employees, a collaborating robot still promotes safety within the workplace but still includes people in the process. The worker is no longer removed from the process; they are merely changed from a laborer to a controller.

An additional, very important but sometimes forgotten aspect of this concept is that SMEs are likely to find themselves with a difficult job when it comes to retraining their staff. In reality, SMEs are probably better equipped to reskill than large corporations. The reason is that they employ a simpler staff, there is better communication, and learning occurs organically within the context of regular work. Most successful changes within an SME, as illustrated in research, find that the employee readily adjusts to changes when training is easy, incremental, and linked to job performance. An employee who can read predictive maintenance notifications is developing a brand-new skill set while still possessing a pre-existing skill set.

It's significant to realize that automation, as well as AI, is generally a job creator rather than a job destroyer. The more intelligent factories become, the more job tasks are generated, such as interpreting data, system integration, basic cyber protection, process optimization, and management of people and machines. In small- to medium-sized businesses, such tasks are not necessarily full-time duties. Instead, they are integrated into existing ones, making individuals in these businesses more flexible with regard to changes, rather than fearful of them.

There is a real problem for small and medium enterprises (SMEs). This is that there are simply not enough people for them to employ. Also, small factories in many areas struggle to find suitable people who are willing to carry out physically demanding, repetitive work. This is where automation and AI can assist, not by replacing people, but by making the process of making things safer, more attractive, and more engaging. Young people are far more interested in small businesses that are technologically savvy than ones that are outdated.

Strategically speaking, the claim that automation is going to steal all the jobs is even more deleterious than the effects of automation. Smaller businesses that shy away from technology out of fear may find themselves at a competitive disadvantage when it comes to other businesses leveraging AI to increase quality, reduce prices, and deliver faster service. In such instances, the job isn't being replaced because of automation—it's being replaced because the business is no longer competitive.

The notion of replacement by automation and AI is an outdated perspective on technology. The fact is that small- to medium-sized businesses are on a path of co-evolution when it comes to the future of manufacturing. It means that humans and computers will develop together, with the strengths of one compensating for the weaknesses of the other. Computers are best suited for complex work, while humans are best at providing meaning.

The most successful small businesses will not be the ones that are most automated, but the ones that assimilate best with the use of technology, using it to extend, rather than diminish, human capacity. The age of automation is not the end of human labor in the factory. It is merely the beginning of a new age, where human labor is more vital than ever before.

3.3 Myth: Artificial Intelligence Is Too Complex for SMEs

It is a common belief within small to medium enterprises (SMEs) that AI is a complicated system to comprehend, implement, and maintain. Additionally, they consider AI as a tool used by researchers, technology giants, and large corporations that employ a team of data scientists. This gives AI a sense of elitism, which leads SMEs to believe that they are not capable of handling AI. This, however, is a misplaced belief because it confuses the hardness of creating AI with the ease of use.

Although the development of AI systems requires a strong background in math, statistics, and computer science, most small and medium enterprises (SMEs), rather than developing the AI solution, find a use for it. An SME manager doesn't need to develop a CNC controller in order to operate a CNC machine, for instance, and similarly, an SME doesn't need to develop a neural network in order to use the power of AI in scheduling, predictive maintenance, or quality assurance. Modern AI systems are designed such that, despite the complexity, an easy-to-use interface is provided, making advanced analyses accessible to anyone.

In recent years, AI has been made more accessible, just as happened with computing, automation, and other technologies that came before it. What once required custom software applications, on-premises infrastructure, and expert staff is now available via cloud infrastructure models, pay-to-play models, and no-code, low-code solutions. Most of the AI software that is used in the manufacturing sector is based on a dashboard that resembles a common spreadsheet/report.

The notion that a perfect, massive amount of data is required for AI is not always true. The quality of the data is of importance, but the latest technologies are capable of performing effectively with imperfect data. Most AI solutions are created with the intention of being used with noisy, incomplete, or inconsistent data, which is a common practice in small- to mid-size businesses (SMEs). In addition, most SMEs are misunderstood in recognizing how valuable the existing data is. For instance, machine logs, maintenance, quality, even operator notes, follow certain patterns that AI is capable of learning from. You don't begin with "big" data; you begin with useful data.

The other source of such a myth is a fear of disruption. The executives of the SME are concerned that with the incorporation of AI, they would need to make massive changes to the existing way of performing tasks, which might result in slowed production. In most cases, the adoption of quality AI in an SME occurs in a subtle manner. The AI software runs in the background, analyzes, and then provides reports without interrupting the daily tasks. In this case, a predictive maintenance application might only notify the user when it detects a strange condition. The equipment continues running, and the user makes decisions on what to do with the information.

From a business perspective, small to medium enterprises (SMEs) believe that they don't have enough staff to support the use of AI. But the fact is, AI reduces the need for human analysis, not increases it. Instead of needing more people to plan and more people to check the quality, what small to medium enterprises need is AI assistance that can enhance what is already being accomplished by their staff. It means that a certain production supervisor can now accomplish a more complicated job thanks to AI assistance.

Industry 5.0 helps in this regard with a focus on human-centric intelligence. AI is not intended to be a solo decision-maker in small businesses. It is a consultative tool that identifies patterns, forecasts outcomes, and provides recommendations, which are later confirmed by human interventions. This is a simpler way of doing things, as humans retain control. The solution doesn't have to be flawless; it merely has to be useful.

The belief that learning AI is difficult is exaggerated. It has been observed that small- to medium-sized business employees quickly learn when they see AI applications used in actual tasks. The more the employees see the benefits of AI, such as how it helps avoid machine failure, reduces rework, or aids planning, the more curious they are going to be rather than apprehensive. The key is relevance, not skills. Training should emphasize how AI can assist with tasks, rather than how AI works.

It is necessary to distinguish overall AI purposes from beneficial uses of AI. Small- and Medium-Sized Businesses (SMEs) do not necessarily require learning factories now. Smaller tasks within AI, such as identifying anomalies, forecasting, or inspecting images, provide most of the benefits with reduced complexities. Such systems operate within bounded domains that are simpler to validate, interpret, and trust. Once such systems are trusted, use of AI by SMEs may gradually increase on their own.

The notion that AI is too complicated is already being challenged by the growing number of vendors who serve small- to mid-size businesses with AI. It is already common for vendors to provide turnkey solutions that come with installation, training, and support. Increasingly, governments, trade associations, and incubators are providing subsidized use of AI solutions, training, and pilot projects for small producers.

It is generally believed that AI is too complex, but mostly, it is a mindset problem, not a technological problem. The small- and mid-size businesses that consider themselves "traditional" and "undigitized" tend to underestimate themselves. In fact, history has shown that such businesses have always managed to cope with technological advancements, such as CNC machines, ERP, or automated inspection systems. AI is the natural next step.

Ultimately, AI doesn't force small- to medium-sized businesses to become tech firms. It's more of a call to become a data-conscious decision-maker. The challenge is not in applying AI but in continuing to make decisions in a more difficult, unpredictable environment with which they are familiar today. In this respect, AI is not a threat but a need—it's a tool that simplifies dealing with complexities.

The brightest, most talented SMEs are not, therefore, the ones with a deep insight into the most profound technical specifics of AI. The brightest are the ones who possess a deep understanding of what is going on within themselves, which AI helps them comprehend. By keeping things simple, dealing with real problems, and collaborating with AI, it is entirely possible to use AI properly. The complexity disappears when AI is treated, therefore, as a useful tool rather than a theoretical concept.

3.4 Myth: Smart Manufacturing Requires Replacing All Existing Machines

The biggest misconception regarding smart manufacturing is that, in order to go smart, one needs to exchange all existing equipment. This is enough for most small- and medium-sized businesses' owners to dismiss even the thought of any changes pertaining to digitization. The machines that are used in the production sector are of huge value, and it has taken decades to amass all such equipment. Even CNC machines, presses, molding machines, furnaces, and assembly lines are not highly advanced, but they are reliable, understood, and paid for in full. This is a result of a lack of understanding of what constitutes "smart" when it comes to manufacturing.

It's not the age of the equipment that matters in smart manufacturing. It has to do with the layer of intelligence on top of that equipment. It's not necessary that a machine is modern; it has to be observable, measurable, and networkable. It's been noticed that most of the successes that have happened in smart manufacturing in SMEs are on machines that are 20, 30, sometimes even 30 years old, because the machines themselves were still accomplishing the mechanical part of the job quite efficiently.

This is commonly known as brownfield transformation. Greenfield factories are different because a greenfield plant is a brand-new plant with state-of-the-art, highly automated systems, which is typically not the case for small businesses in the manufacturing plant sector. Brownfield transformation is a way to upgrade, not replace. It is not a compromise, but rather a sound plan that meets the capabilities of small businesses.

The base for brownfield development is retrofitting. By retrofitting, sensors, data acquisition systems, and communication systems are added to existing machines to enable them to collect data on how they are operating. The addition of such systems doesn't impact the primary function of the machine. Vibration sensors, temperature sensors, current measurement systems, pressure sensors, and cycle counters can be installed on the exterior of machines, with no need for changes in the control systems. Once the data has been acquired, AI software solutions can analyze it to provide valuable information on how machines are performing.

For instance, a mechanical press from the 1990s may lack a digital interface on the machine itself. However, based on the current of the motor, vibrations, and length of cycles, a small- to medium-sized business can identify when a machine is misaligned, worn out, or close to a breakdown. The machine does not change; what changes is the environment around the machine. This is what smart manufacturing is all about.

The other part of the same legend is that smarter manufacturing is all about automation. Automation is a part of a smarter factory but not a necessary part. Most of the smarter manufacturing advantages, such as predicting when maintenance is required, energy saving, and the analysis of production, are only dependent on data. Smaller businesses can derive huge benefits with a complete comprehension of how their machines operate.

Replacing all machines would mean high costs for SMEs. It is also irresponsible from a long-term perspective because a modern machine has a learning curve, along with potential interfacing issues, which result in unpredictable downtimes. The existing machines are already compatible with the particular products, materials, and processes of the company. This is exactly what smart manufacturing honors.

From a financial perspective, the replacement myth ignores the reality of **return on investment**. It has been shown that enhancing with digital components improves ROI faster than full replacement. In regard to a critical piece of equipment, adding sensors and analytics can be done for a fraction of what a full replacement price might be, providing short-term payoffs such as reduced downtime, reduced waste, and improved scheduling precision. For small businesses, such a ROI-oriented approach is not merely advisable, but necessary.

But there is a human side to this. People who run and maintain machines in small businesses know their machines very well. They feel the machines in a way that cannot be described in a manual. If we replace the machines, we might forfeit that knowledge, and people might resist adoption because the new machines feel different. Instead, upgrading the machines people use gives them new skills without taking away what they know. If the operator sees that the vibration measurement is what they have come to know, they'll trust the computer systems.

Industry 5.0 provides a compelling argument against replacing everything all the time. The human-centric approach to manufacturing is not merely a call for periodic updates to technologies but is also linked to sustainability, robustness, and the well-being of human life. Extending the use life of existing machinery with modernization via the use of digital technologies is a sustainability aim in that it reduces waste, locks in carbon embodied in products, and maximizes on past capital outlays.

Smart manufacturing systems are inherently modular. Regarding small- and medium-sized enterprises (SMEs), they are able to choose which machines to digitize when, based on priority, failure rate, or effect of quality. An organization may begin by enhancing the most frequently failing machine, which is later augmented with additional machines once the value is recognized.

The implication here is that in order to think smart manufacturing, you need to throw out everything that means a simplistic all-or-nothing proposition—old vs. new, tradition vs. smart; but, of course, that is never how it happens. What begins happens on a continuum—a factory that is sometimes smart, becoming smarter, becoming very smart, with conditions largely as they were but with a changed means of making decisions concerning equipment.

While the thought that "all machines must eventually be replaced" is a ready-made reason to pause progress, it is actually a hindrance to the vision of "smart manufacturing." When small- to mid-size businesses can observe that their units are no longer an obstacle but an advantage that could become better with data and smarter thinking, their course is clear.

Smart manufacturing is not about ripping up what is there. It is about looking at things in a new way. Machines that were quiet to operate begin to "speak" through data. Processes that were driven by intuition are governed by intelligence. And manufacturing facilities that were once held back by outdated technology discover that their greatest strength is found by integrating the past with the future, not rejecting it.

3.5 Myth: Smart Manufacturing Delivers Instant Results

One of the biggest myths about smart manufacturing is that it delivers overnight, dramatic breakthroughs. This is often the result of marketing, speeches, or rapid success that doesn't portray the work that happened behind the scenes. This is particularly problematic for small to medium businesses. If everyone is expecting things to improve overnight, then real projects that are actually succeeding can appear to be a failure.

A manufacturing process is a complex system that involves both human elements and technology. There are entwined interactions between machines, human labor, materials, suppliers, customers, and other external factors that aren't easily predictable. Smart manufacturing does not alleviate a complex system; it only makes it more visible and more manageable. However, understanding the system is prior to improvement. A certain amount of learning must occur before improvement is possible. It takes time for learning to occur.

A major contributing factor is the expectation of seeing the results instantly. This is the point where the setup of the technology is confused with organizational change. Installing sensors, software, or the AI application is only the initial step. It is necessary to calibrate, provide it with data, test it with real-world data, and then ensure it is interpreted correctly by humans. This is not something that can be achieved instantly, especially for small- to mid-size businesses, as it would mean training with limited resources and learning in a live work situation.

In many effective small- to medium-sized enterprise (SME) case studies, the initial outcome of smart manufacturing, instead of reduced costs or increased production, is often clarity. This is where executives understand precisely where the lost time is, where energy consumption is highest in the machines, or what the continued issues with quality are. This can be an awkward spot for a company because it spotlights areas of inefficiency that were hidden or accepted before.

One more reason that this thinking persists is that the positive effects of smart manufacturing are accumulated over time, but not instantly. An example of this is that a predictive maintenance solution is not likely to completely prevent a failure in the initial week, but it could reduce the number of instances of such breakage significantly after a couple of months. A scheduling solution powered by AI is likely to begin with the same efficiency as that of a human, but as it learns from previous data observations, it eventually starts to perform better than the human.

Patience is important for small- to mid-size businesses since the projects tend to roll out in a step-by-step process. This is unlike big businesses which tend to roll out systems to the entire company at one time. As for small businesses, the pilot is aimed at learning without trying to make quick money back. It is responding to important questions such as Do the data make sense? These answers help pave the way for expansion of rollout processes. Don't look for a quick return on investment here.

It is also important to understand that sometimes smarter manufacturing alters what it means to be more effective before it actually does so. That is, more effective data gathering could indicate more defects or more downtime before it actually does so—not necessarily because it is a worsening situation but because it is measured more effectively. It may seem as if you are moving backwards before you actually are if you aren't mentally prepared for it. It is basically progress. You can't improve what you aren't measuring effectively.

From a human perspective, expecting immediate outcomes is not feasible since behavioral change takes time. Decision-makers who use their own experience should begin using data-informed information. Managers who are accustomed to receiving reports on a weekly basis must become comfortable with monitoring real-time data. Instead of relying on predictably failed systems, maintenance staff could believe their predictive warnings. This type of adjustment demands training, experience, and reinforcement. Smaller businesses that employ multiple staff members in various roles must perform this training concurrently.

Industry 5.0 recommends that change should be gradual and centered on people, not rapid. A company that pushes technology too far is likely to burden employees and damage their trust. Genuine improvement that can last a lifetime is achieved by synchronizing technology deployment with people's acceptance levels. You cannot compel this; you must grow it. Employees would accept the system since they understand its rationale. It would yield fruits by itself.

Research confirms this intuition for long term. Research indicates that while certain applications of smart manufacturing, such as energy or basic condition monitoring, provide rapid benefits, most of the significant benefits occur after the 18-month mark. These benefits include reliable production, reduced variability, more consistent quality, and better decision-making in uncertain contexts. Notably, these benefits provide sustained improvement over reduced costs achieved in short term since they become ingrained in the organization.

Holding a mindset that you'll witness the results instantly is both disappointing and risky—quitting prematurely is risky. Small- to mid-size businesses (SMEs) finding themselves expecting instantaneous results could risk quitting their endeavors just when things are getting started. This would mean that such SMEs could witness a series of failed tests, causing one to question the effectiveness of technology. SMEs with a mindset of learning through smart manufacturing would experience far more benefits.

Ultimately, smart manufacturing is not something you turn on or off. It is something you develop. It begins with observing what is occurring, grows with understanding, and is refined with practice. Each is dependent on what came before it. It is not better or more effective because of reduced costs or additional production; it is learning at a rate of increase that outpaces its competitors.

That promise of seeing the results right away is not valid if one takes a closer look at what a smart manufacturing system is. It takes a long time to work, but it does work. It does not provide major advancements at the start, but it does provide enough advancement that small- and mid-size businesses feel that it is not about massive results but about smarter growth.

3.6 Real Opportunities Hidden Behind the Myths

When you strip away the myths of smart manufacturing, you aren't left with more risk or complexity. It's filled with opportunities, especially for small- to mid-size businesses. Big-picture concepts mentioned earlier often conceal a truth that is refreshingly straightforward: smart manufacturing is for everybody, not just large corporations. It pays off for companies with clarity, agility, and great decision-making, of which small businesses often have abundant quantities.

When SMEs no longer believe that smart manufacturing is for big companies only, they become aware of just how doable modern technology is. This is because small businesses can add sophisticated functionality affordably using sensors, cloud-based analysis, cloud-based AI, or automated solutions that small factories today can use. That's what makes smart manufacturing solutions not a far-off dream that small businesses can apply to their core issues such as equipment availability, reduced quality, lost energy, or poor delivery times.

When people no longer fear loss of employment and comprehend the collaboration between humans and AI, a second chance emerges: empowerment of the workforce. Intelligent manufacturing encourages small- to medium-sized businesses to increase their human workforce's roles, from menial tasks to more analytical tasks. Employees observe machine performance; prediction of maintenance is anticipated before it occurs, while executives make decisions through data flowing in real time, not gut feelings. This enhances productivity, increases worker satisfaction, and helps retain staff, a crucial advantage considering the scarcity of labor in the current labor market.

Getting past the thought that AI is too difficult allows the possibility of smarter organizations. SMEs that employ specialized, small AI solutions notice that things actually get simpler, not more complicated. This is true with respect to planning, finding issues with quality, or dealing with variability. AI does not mean more decisions for SMEs; it means making fewer, better decisions. Being able to work smarter, not harder, is one of the least appreciated aspects of smart manufacturing.

On the other hand, understanding that the current state of machines can be optimized rather than replaced is a vast opportunity for cost saving and sustainability. Small- to medium-sized companies can increase the useful life of existing machines with enhanced digital knowledge, resist the urge to make big-ticket purchases, and reduce their footprint on the planet. Upgrades align with good finance practices as well as Industry 5.0's people-centered innovative approach.

Perhaps the key opportunity that emerges by abandoning the concept of rapid outcomes is the establishment of long-term competitiveness strength. Efficient manufacturing does not promise rapid improvement but delivers progress. When SMEs continue to learn, measure, then improve incrementally, their strength becomes difficult to replicate by their competitors. This eventually results in quicker reactions to any change in the marketplace, stability in operations, or better relations with their clientele.

Taken cumulatively, these opportunities point to one truth: Smart manufacturing is more than a tech vision; it's a way of working. It allows small- to mid-size businesses to compete not only by their size and cost, but by their smarts, their reliability, and their agility. These are important qualities in a world that is increasingly uncertain and constantly shifting.

The true power hidden within the myths is that of confidence, the confidence to begin small, to learn from data, to believe in people and technology not as separate entities but as one, and to plan for growth. These SMEs with a mindset like this quickly realize that with smart manufacturing, they do not become mini-versions of big companies. Instead, it allows them to become better versions of themselves that are not only equipped to face the future of manufacturing but also get to shape it.

Chapter 4
AI-Driven Production and Supply Chain Optimization in SMEs

Introduction Knowing what smart manufacturing truly is and dispelling the misconceptions that exist about it, we finally get to the heart of what is driving the industrial revolution: the artificial intelligence (AI).

While Industry 4.0 enabled machines to interconnect and communicate with each other, AI enables them to think, reason, and make decisions. Smaller firms are looking at this development not only as an important technological feat but as an opportunity to compete with brains, not necessarily brawn.

It allows a small manufacturing unit to sense the small variations occurring around it, predict what is about to happen, and react quicker than would otherwise be possible for human efforts. In manufacturing, for example, it allows for dynamic adjustment of production schedules, detection of defects using computer vision, or anticipation of maintenance needs before something goes wrong. When it comes to the supply chain, it helps predict what the end consumer wants, manage inventories accordingly, or coordinate logistics with a degree of almost real-time precision. What took large planning staffs with expensive software solutions would then be possible through cloud-based platforms with AI solutions.

In the context of small- to medium-sized businesses, AI is not about replacing human resources. It assists humans in smarter work. A production manager with predictions, for example, can plan better, whereas with AI-based buyer forecasts, the purchase process speeds up. This chapter explores the concepts mentioned above by concentrating on what areas of small- to medium-sized manufacturing businesses AI is presently adding real value to production scheduling, predictive maintenance, inventory optimization, demand forecasting, and more. Additionally, it helps you understand the step-by-step implementation of such systems.

A. A. Kadam, *AI-Powered Smart Manufacturing*, SpringerBriefs in Applied Sciences and Technology, https://doi.org/10.1007/978-3-032-21466-9_4

4.1 AI in Production Scheduling and Resource Optimization

Producing a schedule is one of the most challenging yet critical activities in manufacturing. In small- to medium-sized enterprises, it is often the only task that helps in running the business effectively or struggling with never-ending fire-brigade operations. In contrast to large companies that employ planners with formal planning cycles, in SMEs, it is mostly a handful of seniors, experienced individuals like the entrepreneur or production or senior supervisory personnel who manually plan without the use of formal analysis and without strict planning cycles. It is fine for small businesses, but it becomes fragile with more products, varying production orders, and shorter delivery times.

Traditional scheduling systems are too rigid. These systems make plans with the expectation that machines would always be available, materials would always be supplied on time, and priorities would remain the same. This does not work in the small or medium-scale factory setting. A malfunctioning machine, a supplier who is late with delivery, or an urgent order would spoil an entire schedule in a couple of hours. This would mean rescheduling manually with limited information. This could result in inefficient use of machines, frequent resets of machines, missed delivery commitments, or unnecessary frustration for everyone.

Artificial intelligence introduces dynamic data-driven forms of planning into scheduling. This means that instead of using fixed or human-made rules, scheduling software with AI looks at real-time data related to machines, inventories, personnel, or order queues. Such software doesn't implement a schedule once and then forget about it. Rather, it keeps reoptimizing it as things change. This is what helps small- to medium-sized businesses gain the most from using AI.

Optimization algorithms that can test thousands of possible production runs in mere seconds lie at the root of AI scheduling. These algorithms examine constraints such as what machines can or cannot do, processing times, the need for certain tools, material availability, staff work hours, and delivery times. Machines in small- to medium-scale enterprises tend to do multiple tasks, whereas the number of products is large. Thus, it is difficult for humans to tackle such complexities. This is where AI is beneficial, as it does not get tired or get biased by complexities.

For instance, consider a mid-size machining shop that deals with various orders for its customers. Each client would like something else in terms of tolerance levels, production quantities, and delivery dates. While a human planner could allocate orders that are similar to minimize setup time, they could potentially overlook inconsistencies in the end process or resource constraints. A computerized planner considers the entire situation simultaneously, identifying orders which minimize setup times as well as machine utilization rates for delivery dates. If there is an unexpected breakdown of a machine, it instantly identifies alternatives for routing orders to minimize the interruption caused.

Resource optimization is more than just allocating machines. This is also true for human resource optimization. AI-aided scheduling systems also make it possible for us to better utilize human resources. This is more crucial for SMEs, given that they have limited manpower. This is also more relevant for SMEs that are shifting toward Industry 5.0, given that the welfare of human resources is one of their primary concerns.

Usage of energy is another area where AI-based scheduling can provide value. It is a fact that many small- to mid-size businesses use heavy power-consuming machines without knowing the peak times of energy rates. AI can make use of energy rates information and provide advice on shifting non-critical tasks to other times. This eventually cuts down costs without cutting down production. This is difficult to achieve manually but becomes commonplace with the help of AI.

One of the huge advantages of AI scheduling that tends to be overlooked is the role of decision transparency. Today's AI systems do more than scheduling; they provide explanations for why some decisions were reached. It reveals trade-offs, for example, completing quickly versus requiring more setups or favoring one customer over another as a result of contractual agreements. Such aspects of AI scheduling enhance trust between the scheduling planners or operators, allowing them to verify or modify the decisions without necessarily adhering to them. This is crucial in small- to medium-sized companies, where trust is important for the utilization of the system.

When viewed practically, AI scheduling does not necessarily call for the replacement of existing systems. There are solutions for small to medium businesses that work with existing systems such as ERP or MES systems that can be implemented progressively. This means that a business can first implement AI recommendations in combination with manual scheduling, then implement more automation as it becomes more comfortable with the solution. This is appropriate for SME culture since it allows people to learn without disrupting operations.

Technology-based scheduling systems do not reduce human judgment. Instead, they enhance it. Planners no longer engage in repetitive calculations of scheduling, allowing them to engage in big-picture thinking such as talking to customers and looking for ways to improve. In the end, scheduling becomes more proactive, making the organization as a whole better.

Industry 5.0: A human-machine collaboration is what AI scheduling applications mean by Industry 5.0. It means that humans are responsible for values, whereas AI is responsible for dealing with complexity and speed. This collaboration results in a production system that is more efficient, stronger, greener, and more humane.

Production scheduling and resource optimization using AI are something that small- to medium-sized enterprises (SMEs) need badly in order to maintain flexibility in the face of uncertainty. An AI-based scheduling solution helps that SMEs make sensible use of their resources in uncertain situations while continuing to perform reliably. When it comes to maintaining flexibility in order to remain a cut above the competition, a basic requirement for the "smart factory of the future" is production scheduling using AI.

4.2 Predictive Maintenance: From Reactive Repair to Intelligent Prevention

Maintenance work has often been underrated and expensive in manufacturing. It is viewed as a disrupting process for small, as well as medium-scale enterprises. Machines are traditionally repaired after they become faulty or serviced at predetermined times determined by what the manufacturer recommends, as well as what the operator knows. This keeps things running, but it is very inefficient, becoming more expensive with complex production.

Reactive maintenance is where you fix things after you break them. It leads to unplanned downtime, scrap material, untimely deliveries, and stressed-out maintenance staff. Preventive maintenance is scheduling maintenance by time or usage to reduce major breakdowns, often replacing or repairing things that aren't necessarily broken. It is guesswork without facts. Predictive maintenance is information-based, not guesswork.

This is where predictive maintenance comes in. It applies AI algorithms that help predict breakdowns before they occur. It knows what seems normal for each machine by monitoring it continuously for data such as vibration, temperature, sound, current, or the length of a cycle. This allows it to pick out signs of trouble before it becomes noticeable to human eyes.

This means that for small- to mid-size businesses, this feature helps turn maintenance costs from being reactive to proactive. Rather than repairing something that went wrong under urgent conditions, maintenance personnel are given the opportunity to plan for the future. A bearing could be replaced during a scheduled break and not when production is at its peak. A tool could be replaced before it causes waste. A machine could be serviced at a convenient time, not necessarily when it fails.

A lot of small- to medium-sized enterprises (SMEs) feel that some sophisticated machines or numerous instruments are involved in predictive maintenance. However, the truth is that some of the simplest and finest forms of predictive maintenance involve using external sensors that can be attached to the machines you are using presently. Sensors for monitoring vibrations in motor casings, temperature sensors for bearings, or current sensors for power cables are cheap, non-invasive, and very apt for SMEs. Such sensors do not alter the functioning of a machine, which is what SMEs need.

It's not just the sensors that are intelligent; it's the models that interpret the data from these sensors. Pattern recognition is achieved through machine learning engines that reveal wear, alignment issues, imbalance, lubrication issues, or electrical faults. This pattern recognition gets more accurate as it is exposed to real-world data. A predictive maintenance solution is not like other software applications that work with fixed parameters for failure. It knows the individual characteristics of each machine. This is more important for small and medium businesses, where machines are exposed to varied loads and customized processes.

A key advantage of predictive maintenance is that it reduces variability. In small- to mid-size businesses, machines tend to degrade their abilities over time in ways that are difficult to monitor. This results in poorly made products or products that are not consistently produced. This is where predictive maintenance is beneficial by identifying issues before they happen and keeping machines running at peak levels.

From a human perspective, the application of predictive maintenance aligns with Industry 5.0 concepts. This is because AI models do not displace human maintenance personnel. Instead, AI models provide support for maintenance activities. They identify potential issues, provide an estimate of the remaining useful life of assets, and provide the best times for maintenance actions. Human judgment is still involved in determining what actions should be taken, including relying on human judgment for the interpretation of AI recommendations.

A big shift brought about by predictive maintenance is that it alters the way small- to medium-scale businesses gather and spread maintenance knowledge. Traditionally, maintenance knowledge for businesses is locked in the brains of experienced personnel. When such individuals retire or quit their jobs, their accumulated knowledge could be lost. This is where AI-based solutions become handy by using past data or any form of maintenance procedure to gain official or new maintenance knowledge.

In terms of return on investment, some of the fastest returns for companies in the area of smart manufacturing can be achieved in the field of predictive maintenance. It has been observed that there is reduced downtime, maintenance cost, as well as wastage, with many studies observing such results months into the implementation of the solution for small- to medium-sized enterprises with narrow profit margins.

Rollout in small- to medium-sized businesses is typically incremental. A starting point is identifying the key machines whose failure would be the biggest problem. Next is installing sensors, gathering data, building models, then using the predictions of the AI system in conjunction with existing maintenance practices. With time, more weight is given to the predictions of the AI system.

Predictive maintenance is an approach that alters the mindset of small- to medium-sized businesses when it comes to their machines. Machines are no longer passive objects that can break down at any time. Now machines can convey their conditions through data.

In smart manufacturing, predictive maintenance often leads to other major transformations. This is because small- to medium-sized businesses often apply other concepts such as scheduling, quality, energy consumption, or helping with logistics after realizing the power of AI-based predictions for maintenance purposes. It not only prevents breakdowns but also helps enhance confidence in data-based decision-making.

In relation to moving SMEs into Industry 5.0, predictive maintenance is an important means of illustrating the power of technology to enhance human capability rather than diminish it. This is especially important in a world that is characterized by the need for reliability and rapid response times in order to remain competitive, since predictive maintenance is much more than an improvement; it is a requirement.

4.3 AI-Driven Inventory Management: From Guesswork to Intelligent Balance

Inventory management is a delicate function for small and medium enterprises. Inventory is a necessity and a risk at the same time. Insufficient inventory will hinder production, trigger immediate procurement, cause delays in delivery, and disappoint consumers. Excess inventory will tie up capital, occupy storage space, and increase the possibility of obsolescence and damage. To small and medium enterprises with limited financial buffers, a small margin of error can be costly in inventory management.

Traditionally, small- and medium-sized enterprises have used judgment or simple replenishment methods in inventory decisions. This may work in good times but not work well in less predictable demand environments with increased product variety and a heightened level of complexity in supply chains. Manual methods do not have the capacity to incorporate real-world uncertainty such as delays in supply, shifts in demand, product differentiation, and consumption shifts.

Artificial intelligence brings a radical change in inventory management, which shifts from a reactive to a predictive and dynamic system compared to the traditional inventory management system that is simply based on rules. The AI inventory system analyzes historical consumption, production levels, and demand predictions to make decisions on the ideal inventory level based on different circumstances.

For small and medium enterprises, this is a big change because inventory problems do not exist in isolation. Excess inventory in a region can come with inventory shortages in another region. AI is good at identifying these blind spots. Rather than considering all kinds of materials, products, and time series in an organization, AI solutions allow you to see where cash is stuck and where risk is building quietly. Having such a broad perspective on inventory can be very difficult without AI, especially in SMEs with multiple products in small quantities.

One major advantage of using AI-powered inventory management is that it predicts demand shifts rather than responding to them. AI technology takes into consideration seasonal trends, individual customer behavior, and emerging trends which may go unnoticed by people. As soon as a demand begins shifting, inventory recommendations are adjusted in advance, allowing small- and medium-sized enterprises to react to them in a relaxed manner rather than in a rushed manner to make a purchase or arrange immediate shipment.

Supplier behavior is another important factor AI considers in inventory management. Conventional methods are based on standard lead time, but small- and medium-sized enterprises understand this is not a common occurrence. AI analyzes actual performance by suppliers—the frequency of on-time delivery, variability of lead time, and accuracy of fill rates. Suppliers with poor performance translate to increased safety stocks, and good suppliers allow for just-in-time inventory.

Finances-wise, AI-driven inventory management systems improve cash flow, which is a huge consideration for small to medium enterprises. They reduce additional inventory and prevent making last-minute purchases, which gives SMEs extra money to invest in whatever way they see fit. To most small to medium enterprises, inventory optimization can prove to be an equivalent benefit to increasing revenue without exposing the market to many risks.

Importantly, AI does not control inventory decisions. Based on Industry 5.0 concepts, AI acts as an intelligent consultant rather than a controlling device. The inventory planners examine AI recommendations, notice why they were generated, and make decisions based on their own judgment in a given case. For instance, an inventory planner may choose not to follow a recommendation due to an upcoming event such as a sale or a known problem with a specific supplier.

AI inventory management solutions can also improve teamwork in small- and medium-sized businesses. The decisions in inventory management affect other functions in a business, including production, procurement, sales, and finances. However, these functions operate in isolation most of the time. With AI inventory management solutions, all functions can access a common platform with information to make decisions based on an equal level of knowledge.

Smaller establishments will normally have implementation in phases. They will start off with using AI technology on a small but critical set of documents or end products. They will continue to roll out their inventory system to include more products, as they gain confidence in the technology and its effects. The new cloud inventory systems are designed for this approach and have minimal IT requirements.

One of the overlooked advantages of AI-powered inventory management is resilience. AI systems are quick to recalculate inventory risk in case of disruptions such as a shutdown in supplies, delays in transportation, or a sudden shift in demand. Small- to medium-sized businesses can respond in a smart manner rather than react in an inventory risk situation, and this will maintain good customer relations.

Artificial intelligence-driven inventory management removes guesswork from inventory management. In small- and medium-sized businesses, inventory management can function in a less brittle manner, respond without randomness, and grow without getting out of control. Inventory management ceases to be a source of constant worrying, becoming a tool in improving performance and attaining stability.

In smart manufacturing, inventory intelligence relates production to supply chain optimization. Once small- and medium-sized enterprises realize the benefit of predictive and dynamic inventory decisions, they become poised to adopt AI in procurement, logistics, and delivery of the order to customers. What begins with inventory optimization can end up with an entirely connected and smarter operations system.

For small- and medium-sized enterprises engaged in a very unpredictable global economy, AI-driven inventory management solutions go beyond simply reducing inventory. They represent a means of achieving confidence, flexibility, and control. They transform inventory from a passive problem into a tool for intelligent and flexible production systems.

4.4 Demand Forecasting for SMEs

Accuracy in demand forecasting is a critical factor in a production company. Small and medium enterprises do not have a chance to work with advanced forecasting technology because they forecast with an estimate or based on old sales in most cases.

AI enhances forecasting because machine learning can process large amounts of information and identify patterns that would never have crossed human minds. AI considers not just historical sales but other variables such as seasonality, trends, and sometimes social media sentiment.

An example would be a medium-scale producer of furniture employing AI to forecast demand based on search trends and previous sales. As soon as the algorithm recognizes an increased interest in ergonomic chairs, it instructs production managers to make necessary capacity adjustments. Such rapid response enables a competitive edge over other firms and an effective service delivery to clients without increasing inventory.

The AI forecasting systems continue to improve with new information. They become more accurate with time, resulting in reduced waste and a stable cash flow. Small- and medium-sized companies, which in most cases have encountered fluctuating orders, will benefit from this forecast ability, an element that seemed rather challenging to attain in the past.

4.5 Quality Control Through AI and Computer Vision

Keeping product quality constant is a challenge for small- and medium-sized enterprises, and this impacts their reputation. Manual inspection of products involves a lot of labor and can be error-prone, and traditional computer vision inspection systems were a costly proposition in terms of customization. AI computer vision is a radical departure from all this.

With pictures of good products and defective ones, they learn how to detect small flaws. Cameras in a production line snap pictures in real time. The AI immediately categorizes them. Those with defects are marked or removed. Unlike fixed-rule inspections, AI accuracy improves with exposure to more examples, and it can work with different designs and light.

A small packaging firm used an AI vision system for labeling and sealing quality verification. The AI system discovered defects, which were sometimes overlooked by human inspectors, such as a very small misalignment or an insufficient seal. Eventually, customer complaints were reduced by 40%, and inspection expenditures were significantly lowered in a matter of a few months.

Finding defects is just one application of AI, but it can examine quality information from a variety of batches in order to identify the basic causes of variation, linking issues to machines, materials, and people.

Artificial intelligence represents a turning point where the bright ideas of smart factories can be leveraged by all, rather than simply being within reach of large enterprises with deep pockets. For small- and medium-sized enterprises, AI functions as an equalizer where smaller companies can be more efficient, accurate, and progressive in their thinking than before. The common factor behind all highly successful AI projects? They come from a strategy of continuous improvement rather than fear of missing out.

But if a focus on AI as a developing combination of people, technology, and data is considered, a lot can be achieved by SMEs. Later chapters will examine AI solutions and their application not only in a factory setting but in an intelligent supply chain, sustainability, and people plus AI working together. Here begins the move toward a human-centric smart business in Industry 5.0.

Chapter 5
AI Solutions for Smart Manufacturing

Introduction Artificial intelligence is having a rapid impact on manufacturing, and its scope is larger than individual tools and forecasting applications. Chapter 4 discussed artificial intelligence related to scheduling, maintenance, forecasting, and quality. This chapter discusses manufacturing as a whole. Smart manufacturing results when an array of technologies, such as digital twins, IoT, collaborative robots, advanced analytics, and automated manufacturing by AI, comes together.

For small- and medium-sized businesses, it is the integration that is truly revolutionary. Now they can create smart, agile, and intelligent factories, all qualities that were previously accessible only to large corporations that could afford an enormous technology budget. With the advancements in cloud computing, open-source technologies, and sensors, even a smaller workshop can begin to explore and use smart manufacturing solutions that yield tangible results.

This chapter examines how such solutions function, what positive changes they introduce to small- and medium-sized enterprises, and how smaller enterprises can implement such solutions step by step. Rather than viewing AI as an additional functionality, smart manufacturing views AI as an interface that links all entities such as equipment, people, processes, and decisions. Through the ensuing sections, we shall witness how such changes in the use of AI drive the entire process of manufacturing from being an almost entirely manual and reactive approach to being predictive, self-correcting, and human-centric, which is the essence of Industry 5.0.

5.1 Digital Twins: Seeing the Factory Before It Exists

Digital twin: This concept is quite new as far as manufacturing goes, and as a result, many smaller enterprises fail to comprehend it. Basically, a digital twin can be defined as the virtual replica of something real, such as a product, a system, or even

A. A. Kadam, *AI-Powered Smart Manufacturing*, SpringerBriefs in Applied Sciences and Technology, https://doi.org/10.1007/978-3-032-21466-9_5

a whole factory. It does exactly the same as the real one because it also receives periodic data from sensors installed inside it.

Small- and medium-sized businesses are set to benefit greatly from this innovation. So, if a company wanted to improve a manufacturing process in the past, it would do so through a process of trial and error, whereby it would test new parameters, manually swap out equipment, and even shut down the manufacturing process to do so. However, with a digital twin, this process would occur online.

A small sheet-metal shop plans to incorporate a new bending process. Rather than testing various tooling combinations to possibly squander the metallic sheets, the entire process now unfolds in a digital twin. The simulations are provided with the following parameters: where the stress occurs, the amount it will spring back, the amount of energy needed, and the processing time of the process.

Digital twins enhance maintenance processes. When used in predictive analytics, the virtual replicate indicates the parts that are worn out and compares the data to predict failures. This provides small- and medium-sized businesses a powerful tool to take care of the machinery before the problem occurs.

Small and medium enterprises can also implement digital twins by beginning with process twins, which are simulations of how work is done and not of machines. This allows them to prototype modifications to schedules, employees working on what, and where products move before they implement them in actual production. In time, the digital twin can be made more complex by incorporating more data.

Digital twins are more than just a technology—they are a shift in perspective from reacting and observing to predicting and improving, and they function as if they were an intelligent mirror reflecting the SME's operations in ways they were not able to before.

5.2 IoT Sensors and Real-Time Monitoring

Smart manufacturing requires visibility. Without having real-time information about equipment, material, and energy, artificial intelligence does not have much to work with. This explains the significance of the Internet of Things.

The sensors of an Internet of Things, or IoT, system can be compared to eyes and ears of a smart factory. They provide information about temperatures, pressures, vibrations, humidity, cycle time, and human motions. The data collected by these sensors forms the foundation of any decision that an artificial intelligence tool may have to make. Small and medium businesses feel that Internet of Things technologies are difficult to implement and expensive, which isn't true.

Take, for instance, a small company that deals with woodwork. Normally, the technician would have relied on experience in detecting potential issues such as dull tools and an overheating motor. With IoT technology monitoring the level of tool vibration and motor current, potential issues are detected instantly. Notifications are also received on the technician's mobile phone for immediate responses. This translates to less waste, greater accuracy, and increased safety.

The shop floor becomes more transparent by implementing real-time monitoring. The production supervisor can open his/her dashboard to know which machines are operating, which are idle, and which machines are approaching the warning level. The knowledge gained from this functionality assists SMEs to act faster by realigning labor resources, modifying work schedules, or halting bottlenecks to limit further escalation.

Most notably, the IoT enables phased adoption. A small business may begin with one machine or one environment and expand based on their own learning. IoT is not an all-or-nothing solution. Instead, it is small and quantifiable increments toward complete visibility.

5.3 Collaborative Robots (Cobots) and Autonomous Assistance

Industrial robotics can frighten employees and start worries for smaller business owners. However, cobots, or collaborative robots, turn all this on its head. Industrial robots from the past require enclosures and significant space for installation, but cobots are designed for shared working and in the same space as human personnel.

Cobots are useful for tasks such as repeating, boring, or difficult tasks that are hard on the body. The tasks cobots can perform include screw driving, polishing, pallet stacking, picking, or precision gluing. In the case of SMEs, cobots can assist with tasks that are exhausting or cause injuries. This can address the issue of not enough staff.

One of the great benefits of cobots is that it's easy to operate them. It's possible to program some cobots just by moving the arm of the robot along with your hand to demonstrate the action. This is referred to as teaching through demonstration. There is no need for programming skills. Just operate the robot. Your actions will be imitated.

Small-to-medium-scale enterprise, which produced small batches of furniture components, introduced a collaborative robot to sand the edges. Previously, the employees spent a considerable amount of time sanding the edges manually, which led to a number of errors and employee fatigue. However, the collaborative robot was used to complete the repeatable task of sanding the edges, leaving the experts to do the customizing and finishing, which require specialized skills.

Cobots represent the human-centered concept of Industry 5.0. They do not displace human workers but assist and support them, physically and mentally. Smaller enterprises acquire increased speed, enhanced quality, and secure employment without the fear and cost that conventional industrial robots entail.

5.4 Additive Manufacturing (AI-Enhanced 3D Printing)

Additive manufacturing, commonly referred to as 3D printing, has become much more sophisticated. It was once strictly used to build prototype products, and that's not true today. AI enhances this technology by identifying parameters that should be used to avoid defects and by creating structures that cannot be created by human engineers.

In a small-to-medium-scale enterprise, there are many opportunities presented by this technology. Rather than creating prototypes to be made in another firm, they can create those in a few hours in their own firm. Spare parts can also be produced on demand, reducing the storage problem. Clients may also receive special designs without incurring high costs.

AI-based software can also optimize layout by reworking designs to make them lighter and stronger by altering shape and structure. This is highly useful in industries such as automobile production and aviation, where grams increasingly matter.

One small drone manufacturer applied AI-assisted 3D printing for building frames that were 20% lighter than frames processed from machining. There was a substantial reduction in production time and reduced costs, and they were able to test the design in days instead of weeks.

Additive manufacturing is assisting small- and medium-sized enterprises by providing innovation at the shop floor level itself. Through AI-driven modeling and simulation, even companies with limited designers are able to develop sophisticated products that compete with market leaders.

5.5 AI-Driven Process Optimization and Automation

But even beyond tools, AI has the potential to transform the whole process of manufacturing. Process optimization entails analyzing a whole series of variables such as the speed of machines, temperatures, human activity, properties of materials and identifying the combination that results in the highest output with the lowest waste. This has heretofore required highly trained engineers with a great deal of experience. With AI, this can be accomplished in minutes.

For instance, in the injection-molding process, minute variations in pressure or cooling time may drastically impact quality. The AI system trained on previous data can calculate the optimal conditions suitable for each mold, thus reducing the need to experiment.

AI-enabled automation is becoming smarter and far more flexible. Rather than executing fixed instructions, it now reacts to changing inputs. It means it automatically changes speeds, temperatures, and even steps. It is beneficial for small to medium-scale businesses, which use low-volume production.

Automation by AI does not mean robotically completing all tasks by hand but rather enabling processes so that even difficult ones can be handled at a small scale.

5.6 Low-Cost and Open-Source AI Tools for SMEs

AI for the smarter manufacturing sector integrates innovation, feasibility, and simplicity whenever and wherever possible. Things that were possible only through massive systems and the expertise of few are being implemented incrementally and at reduced prices for maximum benefit. For small- and medium-sized businesses, these technologies mean much more than efficiency and cost savings; the change is fundamental.

A factory that can sense, learn, and adapt will become tougher and more reliable. A labor force augmented with AI will become stronger and more able. A firm that predicts, instead of one that simply reacts, will become more competitive, even against larger competitors.

Chapter 6
AI Solutions for Smart Supply Chains

Introduction If the production mechanism can be considered the heartbeat of a manufacturing company, then the supply chain could easily be called the lifeblood. This fulfills the aim of linking suppliers, production facilities, and the distribution channels all under the same umbrella. Small- and medium-sized enterprises had a lot on their plates when it came to managing the challenges posed by the supply chain.

In the traditional model, SMEs are highly dependent on manual forecasting, phone calls, and emails relating to spreadsheet-based supply chain management. Although this would have served them well in the past, the reality is today's supply chain networks are far more complex. Global sourcing, short product life cycles, customized orders, highly unpredictable markets, and consumers calling for quicker delivery demand a faster and more dynamic approach, which cannot be achieved by manual human intervention and judgment. Experiences from pandemics and international tensions have proven this.

But then comes artificial intelligence, and the game changes. Here, AI brings the power of speed, accuracy, foresight, and agility to supply chain management. AI enables small- and medium-sized enterprises to transition from a reactive mode regarding things to a predictive and self-sustaining supply chain that observes changes, analyzes alternatives at the very beginning itself, and changes accordingly with minimum human intervention. Whether demand forecasting or supplier selection, carrier selection, inventory management, or disruption response, AI improves all facets of a supply chain.

This chapter will explore how AI enables small- to medium-sized enterprises to create smart, robust, and transparent supply chains. To serve this purpose, we will discuss practical applications of AI, which include route optimization, risk analysis of suppliers, traceable logistics using blockchain, automated procurement, and optimized logistics. These applications help understand how AI allows SMEs to carry out operations similar to those of larger organizations.

A. A. Kadam, *AI-Powered Smart Manufacturing*, SpringerBriefs in Applied Sciences and Technology, https://doi.org/10.1007/978-3-032-21466-9_6

6.1 AI-Powered Logistics and Route Optimization

The logistics process tends to be the most visible and exposed to risk in the supply chain process as a whole. Small- to medium-sized enterprises' transport costs, reliability, and time to react to a customer are direct determinants of profits and reputation for the business. However, SMEs continue to make logistical decisions with set routes, set times, and intuition alone. Decisions made by experience are adequate when there are no transport risks such as traffic, no fluctuations in fuel prices, no adverse weather, no change of orders at the last minute, or a sufficient number of transport vehicles.

The classic method of route optimization has fixed routes and outdated times of travel. It operates on the assumption that conditions will remain the same throughout the day, but this is not the case in practical implementation. AI-based logistics optimization algorithms eliminate the need for assumptions and do this by incorporating the following factors in real time into the definition of the best routes: traffic conditions, road and weather conditions, weight of the trucks, consumption of fuel, and priorities in deliveries.

For small- and medium-sized businesses, this is even more important because transportation costs are a relatively high overhead cost. A small producer, for example, may only have a few trucks, but inefficient miles, suboptimal vehicle loading, and delivery times can easily hurt bottom-line results. AI-driven route optimization addresses all these inefficiencies by analyzing thousands of possible routes in a few seconds, something even a brilliant human strategist cannot. The end result is shorter routes, less consumption of fuel, better vehicle usage, and timely deliveries.

AI logistics is effective in multi-constraint optimization, which is very important in small- and medium-sized organizations. The path to be taken while delivering products is also not optimized on the basis of distance. It takes many factors into consideration, like time to delivery, which customers to prioritize, vehicle load, working time of drivers, regulations, and even reducing emissions. An example of this is an AI logistics system selecting a path that is slightly longer so it will avoid traffic congestion, which may cause a delay in delivery.

Secondly, AI logistics is also very effective in dealing with changes made at short notice. SMEs, in particular, often face scenarios where they receive urgent orders from clients, where they send products in portions, or where they receive returns. In most of these scenarios, SMEs will have to replan, which means they will disrupt their entire delivery plan. AI logistics can handle changes effectively, which means SMEs will offer improved services without necessarily employing additional human resources.

Fleet management is an important area where AI is applied to optimize logistics. The current technology monitors the health, consumption rates, and driver behavior on a constant basis. Predictive analytics aims to identify early warning signs of increased mileage, poor driving, and improper loading. It is very important for small and medium enterprises, as it increases the lifespan of the vehicle, reduces maintenance expenses, and increases the safety quotient. Even a smaller fleet, when optimized, outperforms a larger, unautomated fleet.

Artificial intelligence in logistics assists humans in making optimal decisions; it does not automate them. Decision makers, including dispatchers and logistics managers, retain control over the final decisions involving talking to the client and managing exceptional circumstances. It assists by providing hints, indicating risks, and justifying how the optimal solution involves a trade-off. This partnership is a perfect example of Industry 5.0.

Sustainability is also an area that is assisted by the use of AI in logistics solutions. Many carbon emissions come from transportation, and small- to medium-sized businesses are under intense pressure from customers and partners to prove that they are being environment-friendly. The carbon footprint feature of some of these solutions helps businesses save money and ensure that their sustainability needs are also fulfilled.

Small- and medium-sized enterprise (SME) implementations are typically simple and phased in. Cloud-based logistics solutions can integrate with existing order management solutions or ERP systems and require minimal computing hardware. SMEs begin with route optimization, scaling to more advanced functionality such as estimated times of delivery, automated customer notifications, and vendor integration when they are ready or need them.

Artificially intelligent logistics means the software solutions helping make their supply chains cheaper also make them more resilient to disruptions in the supply chain. When disruptions arise, whether due to adverse weather, fuel, employees, or infrastructure, AI instantly adjusts delivery routes.

AI logistics ultimately means that transportation becomes a strength rather than a burden. AI logistics gives a small- and medium-sized enterprise improved visibility and control over its delivery network. Smarter optimization replaces rigid planning, and this enables a small- and medium-sized enterprise to better serve its customers and be more competitive without increasing its size and administrative burden.

In intelligent supply chains, SMEs feel the first effects of AI when it comes to logistics. Once delivery times can be accurately forecasted, that leads to more acceptance of AI in sourcing, storage, or demand sensing. Normally, it begins with route optimization, but that becomes the foundation for an intelligent supply chain.

6.2 AI-Driven Supplier Risk Analysis and Decision Intelligence

For small and medium enterprises, the suppliers are more than just suppliers—they are important business partners. Dependability is directly linked to business performance. Compared to large firms, small and medium enterprises depend on fewer suppliers. Dependence is risky. It is true that if one of the suppliers makes a mistake in delivery, quality, or pricing, it can result in increased costs. For small and medium enterprises, managing risks associated with suppliers is not an ancillary problem; rather, it is a primary issue.

Conventionally, the selection of the supplier by small- and medium-scale enterprises happens through experience, relationships, and price considerations. Even now, all those elements apply, but they are insufficient in the ever-changing global supply scenario. The performance of the supplier might vary due to various circumstances beyond the control of the supplier, for example, delivery hassles, labor shortages, regulations, global events, or environmental constraints. It is extremely difficult for human assessment to evaluate those interlinked risks simultaneously. Artificial intelligence introduces a whole new paradigm to the process of assessing and mitigating those risks.

Supplier risk analysis systems using AI are constantly gathering and analyzing data from various sources. These sources of internal data include delivery times, on-time delivery, rates of defects, order-fulfillment accuracy, price changes, and time taken by the supplier to respond to changes. Some of the sources of external data may include prices in the marketplace, delivery time of goods due to transportation, financial indicators, government regulations, weather patterns, and even geopolitical issues.

In the case of small and medium enterprise businesses, this is an important consideration. A reliable supplier up until now may become risky due to certain other issues. AI is able to identify early signs of potential trouble, such as an increase in variance of times or deterioration in the consistency of quality, before major failures occur. This gives small and medium enterprise businesses time to adjust their production plans or increase safety stocks accordingly.

What is good about AI-driven supplier risk analysis is that it provides easy visualization of trade-offs. SMEs always have a hard time trying to make a decision among cost, reliability, and flexibility. They might have a supplier that provides cheap goods and is not that reliable. They might have a supplier that is expensive but is reliable and does not compromise much on margins. AI does not eliminate such trade-offs; it only makes these trade-offs easy to visualize and easy to measure. AI-driven decision-making tools can visualize different scenarios and what the potential effects of such scenarios could be, starting from cost, then inventory, delivery, and finally customer happiness.

AI assists with supplier diversification; therefore, it increases resilience in the supply chain. SMEs find it difficult to diversify because it seems complicated and increases administrative work. AI assists in this area because it continuously monitors other suppliers and determines when diversification of suppliers should take place based on risk levels. Instead of diversifying suppliers for every item, SMEs should diversify just in the risky sectors to retain efficiency while reducing risk levels.

Importantly, supplier analysis by AI technology will not disrupt the traditional supplier relationship. It will make this relationship even better. With clear parameters to measure performance and the ability to share data, communication between SMEs and their suppliers will improve significantly. This will enable SMEs to work with their supplier partners to address problems before they develop into something larger. This kind of collaboration aligns well with the requirements of Industry 5.0.

In terms of human-centered design, AI serves as a support system for making decisions, rather than having autonomous behavior. Procurement professionals remain in command of negotiation, relationship management, and ethical decisions. AI provides insights, points to risks, and presents alternatives; humans make decisions based on context, judgment, and ethical standards. This aided in reducing cognitive overload and helped small- to medium-sized procurement teams, already small in number and overstretched, work with confidence and efficiency.

Often, working toward AI-enabled supplier risk analysis in small- and medium-sized businesses begins incrementally. They start by analyzing a few select suppliers or critical components. Once the quality of information improves and users become more familiar with its application, it is integrated with inventory, production planning, and distribution systems for their various suppliers. Cloud infrastructure ensures such a process starts without incurring significant information technology investments.

In today's environment where the future is becoming ever more uncertain, supplier risk is no exception; rather, it is the norm. Small and medium-sized enterprises that are only dependent on the past and instinct are, in effect, inviting unnecessary interruptions. When supplier risk analysis is powered by artificial intelligence, SMEs benefit from foresight, not just hindsight. Procurement, once an acquiring function, becomes an art that adds stability.

Ultimately, decision intelligence in the area of supplier management means that SMEs can leverage the complexity of larger enterprises and be nimble at the same time. With the aid of AI, SMEs are able to create supply chains that are not only efficient and effective, but also resilient and collaborative, and prepared for the looming unpredictability of the future.

6.3 Blockchain for Supply Chain Transparency

Small- and medium-scale businesses face one of their greatest challenges in maintaining an open and trackable supply chain. If they source raw material from multiple sources, including offshore sources, they can hardly check its origins, ensure its authenticity, or ensure adherence to green initiatives. Consumers and the government are demanding assurance about sustainable sourcing and eco-friendly manufacturing.

Blockchain provides an immutable digital record ledger where all transactions are recorded, timestamped, and cannot be modified. While other databases may be hacked or destroyed with relative ease, blockchain chronicles and verifies each phase from the source material input stage to the final delivery phase.

In the case of small- and medium-scale businesses (SMEs), blockchain technology may appear challenging, but it is easier now with newer platforms. Think of an SME involving food processing and needing to prove its products are compliant with safety standards. Blockchain technology not only records all processes involved, right from sourcing raw materials, cleaning, and finally transporting them, but it also helps an SME trace its exact consignment if it is involved in contamination.

In sectors such as electronics or space, blockchain technology protects the authenticity of components and prevents counterfeit components from entering the production line. Small- and medium-scale enterprises in large supply chains may even require the use of blockchain technology because of the demands of large customers for traceability.

Blockchain increases trust. Rather than the same people looking at everything, the information is accessible to each person. This increases the believability for SMEs, helps with global transactions, and improves the relationship with consumers.

6.4 AI-Powered Procurement and Autonomous Sourcing

Historically, procurement for small to medium enterprises has always remained a largely manual, relationship-driven function. It involves a lot of experience and too much administrative burden. The procurement teams tend to dedicate a lot of time to evaluating quotations, stock levels, price negotiations, and delivery schedules, all of which often tend to be under pressure and with limited information. With increasingly unpredictable supply chains and a shorter product life cycle, this traditional method cannot be effective anymore. AI procurement transforms procurement from a traditional, reactive function to a smart and proactive function.

The key to AI-enabled procurement is to be able to look at a considerable amount of purchase information from different facets. This is where AI systems take into account past prices, purchase quantities, supplier performance, delivery times, and demand forecasts all at once. The system is not limited to contracts based on past performance or a set schedule for a review, searching for the best sourcing alternatives every minute. For instance, if a raw material price is about to escalate or a supplier becomes less reliable with deliveries, alternatives from a different supplier based on different order quantities may be recommended by AI before a situation causes a stop to production.

Autonomous sourcing doesn't exclude the role of human beings in sourcing and procurement. Human judgment is removed from the routine decisions, but it remains an important component. AI helps automate the process of generating purchase orders for some materials when inventory reaches expected levels, thereby relieving buyers of mundane activities. Human beings take care of all other activities, like negotiations for strategic sourcing, while aligning with Industry 5.0, wherein AI supplements human capabilities and not replace them. In SMEs, this is highly efficient and productive for procurement activities since the procurement team might not be substantial.

Another key advantage of AI-enabled procurement is its superior cost management and transparency. AI algorithms identify unforeseen cost-driving factors such as the tendency of making multiple small purchases, differences in prices among suppliers, or unsynchronized ordering. Also, with AI, SMEs can achieve reduced overall procurement expenditure without sacrificing flexibility. In addition, the AI system provides sharp insights that make way for a clear audit path, thus enhancing accountability and ability to address financial requirements.

The practice of using artificial intelligence for purchase decisions in SMEs normally progresses step by step and is risk-free. Most of the cloud tools are easily adaptable to the present ERP or inventory process with minimal installation requirements. SMEs normally adopt the process by using artificial intelligence recommendations together with their traditional approaches to purchase, with increased automation over time once they build trust in artificial intelligence. Also, artificial intelligence-assisted purchase decisions enable SMEs to enjoy stability, speed, and confidence in their purchase decisions, especially during fluctuating market conditions.

6.5 Intelligent Warehouse and Inventory Operations

Warehouse and inventory operations are the most manual and essential processes within the supply chain. Many small and medium enterprises still have not found ways to digitize these processes. It is common in most SMEs to see storage patterns chosen based on habit, inventory processed manually, and pick orders based on one person's expertise. While this might be manageable in enterprises with just a handful of products, with an ever-growing number of products in their portfolios and with faster and more accurate services required by their customers, manual processes reveal their deficiencies.

Warehouse AI systems analyze the way inventory is moving, the frequency of orders, and the constraints of the space to optimize the space. Fast-moving merchandise is placed near the pick area, and slower-moving merchandise is placed in locations that streamline the space. A fixed warehouse would need optimization, but warehouse AI systems optimize themselves depending on the dynamics of the space. This is significant for small- and medium-sized enterprises that have space constraints. This can help the warehouse increase the output capacity without needing additional space.

Order picking and transporting materials within the facility could be applications where AI makes an immediate difference. Intelligent systems optimize picking routes to minimize travel distances and ensure work distributions remain balanced for employees. Now and then, AI may integrate manual pickers with autonomous robots to transport goods between locations and reduce pedestrian activity and manual labor. These applications would be suitable for small- to medium-sized enterprises because such systems could easily be scaled with minimal modifications to accommodate human involvement.

Inventory accuracy is an issue for the warehouses of most small- and mid-scale businesses. This is due to the manual process followed for inventory updating and the late inventory reconciliation, leading to discrepancies between the actual inventory level and the system level. The use of vision technology and inventory tracking technology bridges the gap that occurs due to inventory level monitoring on an ongoing basis. Cameras and scanners keep track of inventory movements and provide early notifications for any discrepancies.

In addition to cleaning up activities, the intelligent warehouse keeps everything stable and aids decision-making. In cases where demand increases, suppliers are late, and product structures vary rapidly, AI analyzes capacity and indicated orders, and recommendations change swiftly. In SMEs, a live view of capacity and decision-making ability is possible because managers receive immediate information about capacity and decision-making, and they can make quicker and more informed decisions. In SMEs, the warehouse is no longer a storage area but a dynamic and intelligent part of the supply chain system.

6.6 AI-Driven Demand Sensing and Market Responsiveness

Uncertainty of demand is a common characteristic of small and medium businesses. Unlike large businesses, which would always have many markets and would also be able to afford a team of forecasters to deal with fluctuations, small businesses would not be able to afford much error margin and would mostly be dependent on a few clients. Not only would it be difficult to notice the sudden change in demand, but it would also create a mess in the production and cash flow of the business if the common methods of demand sensing were used in small businesses.

Demand sensing also operates in a manner that is distinct from other types of forecasting. It employs AI technology to factor in not only historical data of past sales but also consider numerous other real-time variables that indicate how the market is responding at present. Such variables could range from order requests by customers, modifications in their demands, quotes requested, online activity, point-of-purchase input, feedback from distributors, and sometimes other variables that are more peripheral to purchasing, including market and other economic forces.

For small- and medium-sized enterprises, the key advantage of AI-based demand sensing is that of quick response. As the demand begins to emerge, production and procurement people receive early notifications so that they can plan and make use of resources even when the issue does not occur yet. When there is a reduction in demand, the enterprise can reduce production, reduce the flow of inventory, and even safeguard working capital. In high-mix and low-volume environments, where there is great volatility in demand, quick response is greatly appreciated.

Artificial intelligence-based demand sensing assists in coordination between the sales, production, and supply chain functions. In small- and medium-sized enterprises, these three are often performed in distinct departments that do not share any information and communicate slowly. The creation of these AI-based sensing interfaces provides a unified perspective on market activity. This increases collaboration and decision-making across different organizational functions. The sales department can receive accurate delivery dates, production departments will have clear product priorities, and the purchase department can prevent rush deliveries.

Another major advantage offered by demand sensing is its ability to support customer-centric manufacturing, one of the major fundamentals of Industry 5.0. SMEs are thus given an opportunity to align manufacturing strategies with customer behavior in almost real time. SMEs are thus able to respond faster through customizations, drive down lead times, and meet their orders. Customers are thus offered improved services, while SMEs reduce waste generated through producing or maintaining unsuitable inventory.

When applying AI to demand sensing in small- to medium-sized enterprises, it is always done incrementally, and it is easy to get started. The cloud service is also integration-friendly and supports existing IT systems, such as order management, ERP, and CRM. What is more, there is small startup investment. It usually begins by monitoring only a few demand indicators. However, demand sensing is something that, over time, becomes a competitive competence. It increases speed, agility, and market intelligence. It enables small- to medium-sized enterprises to move in response to the market, rather than chasing after it, in a constantly changing market.

This smart manufacturing by AI isn't about making one big leap forward. Instead, it's about making progress that's promoted by sound thinking, by adapting to changing circumstances, and by prioritizing people throughout this journey. Throughout this book, we have observed that small- to medium-sized enterprises aren't weakened by their size; instead, such enterprises derive strength from sticking to their focus. SMEs based on AI can provide themselves with abilities that previously belonged to much larger enterprises because, by making processes easier to understand, identifying threats early on, and orchestrating different decisions throughout different stages of production and within the overall supply chain, SMEs make better decisions with speed, confidence, and purpose through smart manufacturing.

As manufacturing enters Industry 5.0, it is a bright future where smart technology will be complemented by human skills. AI is continually improving in capability, but its greatest power is when it is used collaboratively, computers dealing with complex tasks and humans providing insights, innovation, and values. SMEs that do so will ensure their factories and supply chains are robust, resilient, and responsive to their customers' needs. Smart manufacturing is not about changing SMEs, but about making SMEs better at doing what they do best: innovate, be nimble, and be a crucial part of the global industrial ecosystem.

Chapter 7
Practical AI Solutions for Everyday SME Manufacturing Challenges

Introduction SMEs have very challenging realities when it comes to manufacturing. SMEs are faced with daily decision-making with tight timelines and resources. They can easily be affected by even the smallest changes. Unlike the bigger organizations, SME organizations lack the specialized team members, the big enterprise software, and the long planning horizons that are needed to deal with uncertainty. SME organizations require appropriate, actionable, and easily implementable answers.

This chapter moves from theoretical constructs of digital transformation to real-world solutions and implementations of solutions for problems faced by SMEs. The solutions presented in this chapter will be specific AI solutions designed for specific problems in SMEs and will be built upon dashboards and decision-support systems that help to extract valuable insights from existing data without complicating things. This will help SMEs to extract insights and implement them without increasing complexity when faced with current technological problems.

The emphasis of this chapter is on the way that solutions are developed, not just the technology employed. Every problem is broken down into its underlying issues—then a practical, step-by-step analysis of the way data, AI models, and dashboards interact. Genuine examples are used to illustrate the way that SME businesses can get started, leveraging the technology they may currently employ, and then build up the capabilities. These examples are not aimed at full automation but at the way that human intelligence enhances judgment.

Industry 5.0 sees a transformation in the methods SMEs need to adopt regarding smart manufacturing. AI is not a replacement for experience, skills, or knowledge, but a companion that assists in pointing out problems earlier, making faster decisions, and acting more boldly. Through the end of this chapter, you will have a clear understanding of how effective AI solutions can be built, implemented, and scaled, which shifts operations from being a daily challenge to a resilience, stability, and growth opportunity.

A. A. Kadam, *AI-Powered Smart Manufacturing*, SpringerBriefs in Applied Sciences and Technology, https://doi.org/10.1007/978-3-032-21466-9_7

7.1 Problem 1: Limited Inventory Visibility and Working Capital Constraints in SMEs

Inventory problems are one of the most frequent and expensive problems faced by small- and medium-sized enterprises. Most SMEs find that their inventory is composed of the biggest chunk of their working capital, even as their inventory management practices remain old and antiquated, with most of the work done manually and in piecemeal fashion. The inventory is typically tracked by using spreadsheets or ERP reports from time to time or simply observing the ground conditions on the shop floor.

The conventional inventory management system is unable to meet the increasing amount of data, greater product choices, and unstable supply chain. The scenario in which small- and medium-sized businesses often find themselves is bizarre: their warehouses are full, and yet their production units run out of stock. The slow-moving items are overstocked in inventory, thus tying down funds and space, and crucial parts are below safe levels without any prior notice.

The issue isn't the lack of effort and skill. The problem is the missing integrated and real-time/predictive inventory knowledge. Static reorder points and historical averages assume that the future will be the same, and that's just not how SME environments work.

7.1.1 Solution: AI-Enabled Inventory Management Dashboard Using Power BI

The solution for this, which has practical applications for small- and medium-sized businesses, involves the creation of an automated and data-driven inventory dashboard using Power BI, which was based on Industry 4.0 concepts. Rather than having inventory that remains constant, this system views it as something that keeps fluctuating and therefore needs to be monitored, predicted, and reacted to.

The key to the solution is the integration of several streams of data related to the inventory into an analytics solution. The data that is imported into the solution includes the part number, the daily usage (demand calculated from annual usage), the number of the item that is available and in stock, the owner of the inventory, and the calculated critical inventory levels using AI.

Unlike in traditional reports, this Power BI dashboard solution, instead of only displaying stock levels, applies descriptive as well as predictive analytics in categorizing inventory into safe, warning, and critical levels. The critical levels of inventory are predicted through past usage and future demand. It is proactive rather than reactive in terms of inventory management.

7.1.2 *How the Dashboard Works in Practice*

This solution is effective because it is user-friendly. The data from the inventory, whether from Excel files, ERP databases, or SQL databases, is cleansed with the use of Power Query, such that the data is accurate before analysis. Relationships between the data sources can be established in the data model provided by Power BI.

The dashboard, when in operation, displays the part numbers that are most likely to fall below critical levels, rather than the whole inventory. This method directs human attention toward the areas that require attention. The study reveals that the dashboard displays the critical level of the quantity based on AI calculations, making the risks conspicuous to the inventory control engineers and managers.

One of the major characteristics of Industry 4.0 in this solution is automation. One of the ways this solution works automatically is by warning the engineers concerned when the stock level reaches a critical point on a daily basis. This warning goes out automatically, with no human labor involved, in order to ensure risk distribution.

7.2 Problem 2: Unstable Production Schedules and Missed Delivery Commitments in SMEs

Scheduling in small and medium enterprises is never calm and predictable. Schedules in most SMEs begin with the best possible information available. Yet the information changes frequently as the day progresses. Machine breakdowns, late deliveries of materials, reworks, lack of labor resources, and customer requests disrupt all that has been planned. Scheduling changes from planning to firefighting.

Such an instability raises huge problems. New priorities keep being allotted to the operators without much detail, too many changes in the machine setup are required, and the workload keeps piling up in an uneven manner in the workshop. A great deal of the management time is consumed by firefighting, and the production manager has a tough time committing delivery commitments to the customers. In a considerable number of SMEs, untimely deliveries are not because of a lack of capacity but a lack of coordination.

This isn't a case of bad planning. The issue is that we work within a fixed schedule in a dynamic environment. A traditional schedule relies on the premise that all machine resources are always available, that all material will arrive on time, and all priorities will remain the same. If one of those conditions isn't met, the entire schedule is no longer useful. This is especially important in the small and medium enterprise community where they lack the ability to make real-time decisions.

What small- to medium-sized businesses need, rather than a timetable that might be optimistically planned, is a system that changes with events as things happen. This system must help in making decisions, not make decisions itself. It has to point out problems ahead, rather than at the end, with clear guidelines on how to plan during disruptions.

7.2.1 Solution: Smart Production Monitoring and Scheduling Decision Dashboard

The way to address this is through a smart dashboard that monitors and manages production scheduling. This is because this technology connects data from the shop floor to AI decision-support systems in manufacturing. The concept here is that this technology creates a "live" system when it comes to matching demand and supply, as opposed to having one plan per day.

Fundamentally, the value of the dashboard is to consolidate the production data, namely work orders, routing, cycle times, setup times, machine availability, and due dates, in one place. This alone is of great value as it provides a common viewing point as opposed to spreadsheet views and conversations. The key here is when the analytics models analyze this data to identify problems, predict issues, and compare scenarios for the schedules.

The AI component does not attempt to solve a schedule that is mathematically "optimal." In the small- to medium-sized employers (SME) environment in which a small business operates, these types of plans are not always feasible. A lot changes in this environment. The program relies on risk priorities. It is always looking at which orders are potentially going to miss their deadlines, which machines are becoming overworked, and where in the system issues may propagate. This allows managers to get their intervention underway while still having options.

The dashboard is designed for exception handling. Rather than requiring the user to investigate each work order, the dashboard reveals circumstances for which attention is required. This reduces cognitive overhead and ensures that precious human resources are utilized for the maximum impact.

7.2.2 How the Scheduling Intelligence Is Formulated

The scheduling intelligence analyzes the data from several perspectives simultaneously. Time-oriented indicators look at how much buffer time exists between remaining processing time and due dates. Resource-oriented indicators watch the use of resources such as machines, queues, and bottlenecks. The disruption-oriented indicators include historical downtime data, availability notifications, and known capacities such as labor skills and tooling availability.

These factors are used by the models to generate a delivery risk for each active work order. This delivery risk changes with time. Based on the progression of the work and when machines start and stop, or when there is a change in the priority level, the associated risks are automatically adjusted. Instead of wondering which order will be late, the system provides that information.

The dashboard provides action recommendations and not commands. These may be rescheduling jobs in different order priorities, transferring jobs to other machines, deferring the processing of low-priority jobs, and even performing minimal overtime if it reduces the risk of delivery. All action plans provided include an explanatory comment on what is involved.

7.2.3 Practical Example: High-Mix SME with Daily Scheduling Disruptions

Consider there is a small manufacturing company whose workshop contains two CNC milling machines, a CNC turning machine, and an assembly point. The company carries out the production of custom-made subassemblies, which have different routes with occasional changes in priorities. Every morning, they plan their production schedule, but sometimes unexpected things happen.

Normally, two types of work orders are in process on a typical day. The first is a high-priority customer order with a strict time constraint that has several steps in milling, turning, and assembly. The other is a low-priority internal stock request with a flexible time constraint. Midway through the day, one of the milling machines breaks down.

In the conventional arrangement, the supervisor reassigns the jobs by priority and intuition. The highest priority task goes to the second milling machine, but the impacts on turning and assembling are uncertain. Later, the assembling process sometimes runs out of components; overtime requests are made unnecessarily, and the level of confidence in delivery becomes low.

It means that because of the smart scheduling dashboard, there is an immediate response. It rapidly identifies which of the work orders are affected by the machine shutdown and then recalculates how much capacity is remaining on the other machines that are still available. It reveals that rescheduling the high-precedence order on the second milling machine maintains delivery on course provided that such rescheduling happens within a certain time limit. It further reveals that the impact on overall performance associated with postponing shipment of the internal stock order is minimal.

The supervisor checks the dashboard and agrees with the recommendation. The operators are informed of the priorities. There is no overtime requirement. The assembly process remains balanced. The customer order is on time. This decision and result are saved in the system. This will improve future recommendations.

7.3 Problem 3: Unexpected Machine Breakdowns and Reactive Maintenance in SMEs

Machine breakdowns in small and medium manufacturing enterprises not only translate to technical issues but can also be production crises in terms of impacting the production schedule, labor, utilization of inventory, and meeting customer deadlines because one machine breakdown can shut down an entire production line in small manufacturing enterprises or SMEs. However, maintenance in SMEs is generally reactive, as machine repairs occur only after machine breakdowns.

Despite the predictive routine maintenance, it is typically determined by fixed time intervals or broad guidelines from the manufacturer. Of course, it goes a long way in preventing huge problems, but not exactly the way the equipment is being utilized in small- or medium-sized enterprises. The equipment may be working in different cycles, using different materials, and so on. Because of these variations, certain components might be maintained prematurely—which means wasted resources—and others may fail because the wear cannot be detected.

An important characteristic of the problem in the SME is the concentration of knowledge. It usually takes a handful of experienced technicians to truly understand by intuition the behavior of machines, due to their abundant experience. It becomes problematic if these people are not present, and maintenance can be done only with the normal delay of problem solving. All the important knowledge is not documented; therefore, the maintenance process is highly dependent on who is present in the office.

The challenge with traditional maintenance solutions isn't that they lack effort or expertise. The issue lies in the early detection of machine condition. Regular maintenance approaches usually detect issues only after a machine malfunction. This means that the solutions are usually limited at this point, resulting in downtime as well as escalating costs. Small- and medium-scale organizations require early identification and interpretation of changes in condition with plans that prevent interruptions.

7.3.1 Solution: Predictive Maintenance and Machine Health Dashboard

One way to address the issues associated with maintenance would be to ensure that the machines make use of a predictive maintenance and machine health dashboard that monitors equipment at all times and converts their information into insights. Such a system would monitor equipment based on sensor information and AI to identify indicators of equipment malfunction.

The machine considers machines as active sources of data and not just as assets. It monitors crucial indicators such as vibrations, temperature, current draw from the motor, as well as the stability process. These indicators help the machine learn the norm. Once the norm has been set, any anomaly becomes apparent early enough before the operators and the maintenance team.

The system is designed for decision-support purposes and not automatic control. It does not automatically stop the machinery or schedule maintenance. It simply provides the maintenance personnel with information on which machines are healthy, which one is deteriorating, and which one requires maintenance soon. This ensures that SMEs perform maintenance actions that suit the time scheduled for production.

7.3.2 How the Maintenance Intelligence Is Formulated

The intelligent piece of the dashboard occurs when condition information is combined with a clear understanding of the operation of the machine. Rather than being presented with information from sensors, the AI reads the information in the context of factors such as the load on the machine and the runtime.

Machine health comes through as easily interpretable indicators such as system health scores, anomaly levels, and risk windows. Instead of forcing the tech to interpret the graphs, the dashboard will tell the user the status of the machine in terms that are easily understandable—healthy, warning, critical, with explanations that associate the pattern on the graphs to a possible physical cause, just like the way small- to medium-sized maintenance teams think.

Meanwhile, the results of the maintenance are documented. Eventually, the system becomes able to know what patterns have caused the failure in reality and what have not, hence increasing accuracy without relying on the experience of one individual in the organization.

7.3.3 Practical Example: CNC Milling Machine Predictive Maintenance

Take a small to medium enterprise where CNC milling machines are used. Such machines occupy a prominent place in their production line. Often, they stop running unexpectedly as a result of issues such as spindle, axis drives, and overheating when performing heavy cuts. Usually, such issues would only be addressed when one could hear vibrations, see the surface finish change, or when warning lights turn on, meaning that the condition will have worsened.

To implement the predictive maintenance process, the small-scale manufacturer installs a simple but useful sensor arrangement on the CNC machine. A vibration sensor is mounted on the spindle case close to the bearings to detect mechanical signatures. A temperature sensor monitors the spindle case temperature to identify unusual heating. A motor current sensor is placed in the electrical booth to monitor the spindle load condition in similar tasks. A coolant flow sensor is placed where it can monitor the coolant flow rate to the cutting area to provide early indications of issues with the pumps and filters.

Over several weeks, the system learns what normal behavior is when cutting under different conditions. The system knows what normal levels of vibration, current, and temperatures are during normal machining. Once this is understood, deviations are apparent. Vibration at a slowly increasing level in certain frequencies, accompanied by increased levels of current and reduced cooling rates when the cut is completed, points to spindle bearings that are failing. Reduced levels of coolant, detected before any warnings, are why temperatures are elevated during prolonged cuts.

The following results are displayed on the dashboard. The results indicate that the condition code for the spindle improves from healthy to warning on the same topic, since the vibration and temperature trends correlate more to bearing damage than to issues related to the tools. The system predicts a time frame for the maintenance to be conducted before the risk becomes critical. Maintenance personnel act proactively on the situation and perform the replacement and/or servicing before the equipment fails. This enables the SME to prevent unexpected downtime, preserve the surface finish integrity, and reduce the workload on the staff.

7.4 Problem 4: Supplier Delays, Procurement Uncertainty, and Material Risk in SMEs

In the case of small- and medium-scale manufacturers, the efficacy of the supplier network often becomes the determining factor for the smooth operation of the production system or its frequent shutdowns. Unlike large manufacturers, SMEs cannot afford the luxury of having several supplier channels and risk analysis personnel; instead, they often deal with a limited set of supplier companies for critical components, bought-out parts, and subassemblies.

In most small and medium enterprises, purchasing decisions continue to depend on past relationships, quoted lead times, and unit item prices. As important as these factors are, they fail to indicate how supplier behavior is actually changing. Delivery disruptions typically become visible only after a supplier is late, requiring a small to medium enterprise to take reactive steps such as expediting their ship, adjusting manufacturing schedules, and negotiating with customers.

The situation is compounded by the fact that the procurement process affects a host of other activities in an enterprise. When an item arrives late, it affects not only the procurement team but also the whole enterprise's inventory and production schedule, the labor utilized, and the commitment to the end-customer. Without a mechanism to assess the risk of suppliers and products early on, SMEs make judgments on their own without a view to the proximity to a crisis. The point to note is that suppliers are not working to fail but lack early visibility.

7.4.1 *Solution: AI-Based Supplier Risk and Procurement Intelligence Dashboard*

One possible remedy for procurement uncertainties is to use an AI dashboard for supplier risk and procurement intelligence. This tool is always observing what suppliers are doing, and in turn, what they're doing in operations. Rather than considering a procurement option purely from a transactional perspective, it becomes a decision-making tool for anticipating risks and taking preemptive action.

To begin with, the dashboard in essence centralizes information on purchasing, quality, and stock. Thus, the performance of the suppliers in terms of, say, delivery reliability, stable lead times, quality performance, or price, is not checked intermittently or by guesswork. Rather, their performance on a daily basis helps SMEs identify changes in trends with regard to delivery reliability, stable lead times, the outcome on quality, or price. AI models analyze this information in an effort to identify tendencies prior to significant shortages.

The system does not attempt to make supplier selection decisions or usurp the role of the buyer's power. Rather, it displays which suppliers and which products pose high-risk scenarios, why the risks associated with these scenarios may be increasing, and what implications the risks may carry within the production and inventory environment.

7.4.2 *How Procurement Intelligence Is Formulated*

Supplier risk intelligence can be derived by taking the unrefined data of procurement and using it to create meaningful indicators of behavior. Purchase order history provides information about true lead times, shipments, and the tempo of response to change. Quality inspection information provides insight into the trend of defects. Price history information provides insight into the swings of price and exposure to costs. Taken individually, they all are meaningful; together, they form a moving picture.

These factors are used by the AI models in the creation of an ever-updated risk profile for suppliers. Variability, as opposed to averages, is pointed out by the system. A company with consistent average lead time but rising variability would be ranked as more at risk compared to a competitor with slightly longer but consistent lead time. Also, small issues in the quality aspect rising gradually would be an indicator prior to large defects.

This intelligence is quite valuable when connected with actual operations. The dashboard correlates supplier risk with the amount of inventory on hand, the value of inventory, and production requirements. Rather than asking whether a supplier is "good" or "bad," the system responds to a more valuable question: What is the likelihood that this supplier will impair production within the current planning cycle?

7.4.3 *Practical Example: Managing Risk for Two Critical Suppliers*

Suppose there's a small to medium-scale enterprise that manufactures some industrial equipment using two/components requiring procurement from vendors: machined aluminum housing sourced within the country and an electrical connector subassembly sourced from abroad. Both components are required to complete the manufacture of the equipment, for which there are no readily available alternatives.

In the past, the local supplier has been very reliable with respect to delivery time and defect-free delivery. Foreign suppliers were less expensive with longer delivery times. Currently, there have been occasional instances of delays experienced by the SME from the foreign supplier. However, these issues were resolved after they arose through expediting and rescheduling.

The supplier intelligence dashboard indicates that supply chain data is constantly compared across different perspectives. The data points out that overall lead time from the foreign supplier remains fairly stable, though the variability is up, while there is also a trend of late deliveries and slightly increased minor defects during quality inspections in this regard. Meanwhile, this is contrasted by a decrease in supply chain coverage related to the connector subassembly because of increased customer demand.

This is highlighted as rising material risk on the dashboard. This is represented as: If the next shipment is more than a week late, it will disrupt production in the next planning cycle. This risk is well explained by more variable lead times, less inventory coverage, and shifts in quality trends.

Chapter 8
Future of Smart Factories and Industry 5.0 in SMEs

Introduction The world of manufacturing is at a tipping point. Technology is advancing rapidly, and changing social demands, sustainability, and human ingenuity are also evolving. Industries 4.0 were about automation, digitalization, and interconnectivity. The emerging Industries 5.0 are much more focused on people. They look to intelligent factories that merge what technology does best with what humans do best: their adaptability, empathy, and problem-solving abilities.

In the case of small- and medium-sized enterprises, this paradigm shift is no longer a trend but an opportunity to redefine themselves in a new global setting with speed and new ideas having as much importance as scale. Here, unlike huge enterprises with very rigid structures and processes, small- and medium-sized enterprises have the freedom and flexibility to shift to collaborative robots, artificial assistants, digital twins, and sustainable processes faster and in a more innovative and customized manner.

"The future of manufacturing is no longer a cold environment with a great amount of automation that replaces people; it's a warm and adapting system that leverages people and automation together and allows them to do what they do best." In this chapter, we will explore what the future holds and how small and medium businesses can become the next-gen smart factory and why Industry 5.0 might very well prove to be the friendliest change for SMEs.

8.1 The Shift from Automation to Human-Centered Innovation

The initial stages of the automation industry focused on one thing: machine efficiency. Machines were introduced wherever manual labor was possible, reducing gaps, and enhancing production. Industry 4.0 escalated the idea, connecting various

A. A. Kadam, *AI-Powered Smart Manufacturing*, SpringerBriefs in Applied Sciences and Technology, https://doi.org/10.1007/978-3-032-21466-9_8

machines, systems, and data sources in highly automated, cyber-physical systems. This move had definitely enhanced the aspects of productivity and transparency, but one important learning was that even then, efficiency, resilience, sustainability, or competitiveness is not guaranteed.

Industry 5.0 is born from this concept. It doesn't reject automation but gives it a different purpose. The objective is no longer to automate people but to employ automation for and with people. Under this philosophy, technology is no longer the driving force behind manufacturing but, instead, it is the human. The intelligent factory of the future is designed not to diminish but to enhance human capability, that is, using AI, robots, and information technology to enhance what people are best at: things such as creativity, practicality, morality, and flexibility.

This change means a lot to small and medium enterprises (SMEs) because small manufacturing enterprises rely greatly on the skills of people who understand the task at hand, operators with a feel for how machines react under different circumstances, managers with a workaround solution for unsolved problems from their own experience, and managers with a set of priorities with which they can only imperfectly match data. Fully autonomous systems can be useless in this context since the working practices of SMEs are not typically standardized or predictive. Industry 5.0 argues that the adaptability of human workers is something to be valued rather than a weakness to be discarded.

Human-centered innovation is about designing systems to fit, not the other way around. This would include AI systems that explain decisions made instead of following a secret command, universal robots for shared work with a human, and technology systems that aid decision-making instead of overwhelming with information. What this might look like in a small- and medium-sized factory is an AI that alerts a supervisor to a growing quality problem and recommends a solution, but leaves it for a human to decide.

There also comes a new approach to thinking in terms of human tasks. Industry 4.0 tried to automate all tasks away. This made people very concerned about their employment as well as their skill development. Industry 5.0 brings into focus the concept of sharing tasks rather than removing them. While robots will do the tasks that are repetitive, risky, or difficult to perform, tasks involving monitoring, problem-solving, managing, or optimizing will be handled by human beings. This will greatly enhance productivity as well as employment in smaller companies experiencing a lack of employment as well as skill development.

Human-driven innovation also resonates well with sustainability and social responsibility as core aspects of Industry 5.0. There is greater pressure for small- to medium-sized enterprises to be responsive to the issues of ethics in business, people and the workplace, and the planet. Technologies that can extend the life of machines, reduce waste, minimize energy consumption, and provide a safer working environment are very useful in both the short-term and long-term aspects of broader societal goals. These outcomes can be achieved not by increasing automation but by improving the interface between people and machines/data.

Strategically, the adoption of human-centered innovation allows SMEs to compete on aspects that large corporations find challenging to cope with. The advantage that large corporations have over SMEs is size, whereas SMEs are able to remain closer to their employees and their operations. Technology used in Industry 5.0 further enables this by offering insights without reducing the aspect of human judgment.

But a shift from automation efficiency to human innovation represents a level of maturity that can be seen among intelligent manufacturers. This implies that optimal factories are not those that employ the least number of personnel but those that develop along with technological growth. For small- and medium-sized manufacturers, it's not a choice to remain relevant but an imperative that ensures the value accorded to intelligent manufacturing also acknowledges its humanity.

8.2 Collaborative Robots (Cobots): Redefining Human–Machine Partnership in SMEs

Robots were always envisioned in large, costly manufacturing plants, behind barriers for safety, with repetitive and boring tasks performed at incredible speed, and maintained with specialized technicians. To many small- to medium-sized businesses, robots were thus always unaffordable and unsuitable for use in dynamic, human-centered processes. Collaborative robots, or cobots, are new in this regard. They were designed to be used alongside other humans in achieving tasks and represent Industry 5.0's pronounced emphasis on human beings.

Cobots are different from traditional industrial robots. This is because cobots are smaller and lighter. In addition to this, cobots have a substantial number of sensors. These characteristics make it possible for cobots to work along with humans. It is important to note here that cobots are designed in a manner where they work collaboratively with humans. In the case of small businesses, where the products vary frequently and the quantity is also low, cobots are more useful compared to automated robotic cells.

In the case of small- to medium-sized businesses, the benefit of using cobots is that they eliminate physical effort but also retain control over the process with the help of humans. Some ergonomic issues exist with many small manufacturing enterprises concerning repetitive lifting, awkward postures, a lot of vibration exposure, or tedious manual tasks. Cobots can efficiently undertake the difficult or dull tasks associated with manufacturing, such as placing parts into the product for assembly, pallet stacking, screw driving, polishing, or operating machines, while the human does the examination and adjustment tasks.

Another key advantage of cobots in small- and medium-sized enterprises is flexibility. Conventional automation systems demand rigid layout configuration, long programming cycles, and regular product characteristics, which are hardly found in small-scale manufacturing. Cobots, however, can be easily reconfigured in short cycles, sometimes through GUI software or even manually to guide the robot through specific tasks. This allows easy reassignment of different tasks to the same cobot. A single cobot can be used for assembly in 1 week and for packing in another.

In a labor perspective, cobots are used to help people develop new skills, not lose their jobs. Workers who were only doing manual labor are now placed in a position to supervise the collaborative processes. They learn to customize, interpret their environment, and improve their processes. Industry 5.0 is about doing valuable work and learning new things. The utilization of cobots in midsize companies has proved to increase worker engagement as it reduces fatigue, allowing workers to use their expertise in a more valuable way.

Cobots also alleviate labor shortages; this is a great concern for SMBs. The truth is that fewer employees are interested in carrying out such repetitive or strenuous manual labor. Cobots bridge this labor gap without leaving anyone redundant. The technology enables SMBs to produce more using their current employees in a more effective way instead of relying on new employees. This is an advantage to SMBs in cases where expertise is scarce.

On the economic side, cobots require simpler procurement compared to classical robots. Cobots also require simpler setup, as they can easily pay back themselves after 1 year since they increase production, decrease rework, and lower injuries and sick leave. Because a cobot is able to work side by side with current equipment, it easily integrates into current facilities.

The key to the success of cobots goes beyond the technology to the thoughts of an organization. In small- and medium-sized enterprises where the primary approach toward cobots is as a means for collaborative working rather than a cost-cutting solution, the greatest benefit is likely to be reaped. When employees realize the value of cobots and come to believe that they are there to help them rather than replace them, speed and innovation begin to increase.

In the future intelligent factories, cobots represent a significant paradigm shift in terms of our thinking about manufacturing. This sector is actually proving that automation can be flexible, not isolating or brutal. Industry 5.0 might be a dream for SMEs, but collaborative robots provide a very clear and people-centered path forward, where increased productivity and respect for people can happen at the same time.

8.3 Hyper-Personalization and Agile Production in the Industry 5.0 Era

The future of manufacturing has more to do with the person and less to do with the average consumer. As markets continue to mature and the level of competition expands, the ability to build many similar products simply won't cut it. Instead, consumers have come to expect products that align with their needs, interests, and values whether that relates to their performance, appearance, sustainability, or the speed at which they can be delivered. The trend of personalizing products heralds a different era for factories, and SMEs have a remarkable advantage.

Conventional mass production is based on standardized components and large-series production. It is best suited to large businesses that cater to one large market, but where there is diverse and unpredictable demand, it is not as good. Small- and medium-sized businesses (SMEs) are inherently closer to their customers, as they may involve themselves in custom work, small series, and design modification. Industry 5.0 technology enhances this even further as it facilitates a production system that is able to produce customized products while maintaining efficiencies.

Hyper-personalization is achieved when design systems powered by AI, flexible manufacturing systems, and real-time data come together as one decision platform. The AI system examines customer inputs, such as design parameters, order history, and usage behavior, to generate product variations. Instead of having to engineer each customer variation manually, small and medium enterprises find it feasible to apply intelligent design rules that enable dynamic scaling, material, or feature modifications on the fly, thus significantly reducing the amount of work related to engineering and opening up their product offerings.

Agile production is not only about product design but also about production in the plant, according to Siemens. Intelligent scheduling, modularization, and flexible automation enable smaller production companies to efficiently produce multiple versions of a product one after another with minimal downtime. AI-based scheduling algorithms place orders in a way to cut down on production time but also take into account the customer's demands for product customization. Collaborative robots are ideal in those flexible operations such as assembly, painting, or final product finishing where a lot of variation is not a problem. Human intuition, skill, and quality control are also at the core of personalized manufacturing.

Hyper-personalization is a great advantage for small- to medium-sized businesses because it alters the way a company interacts with a customer. Customers will perceive the company as partners instead of just a vendor if the product meets the needs of the individual. The company will receive responses faster, and customers will be less concerned about the cost of the product. AI technology assists the company to analyze this response and make improvements based on it, which is also beneficial for Industry 5.0, which stresses the creation of value for the individual.

Operations-wise, agile and customized production means less waste. By producing items that approximate actual demand, there is less overproduction and additional inventory. Resources are maximized, and obsolescence is lessened. For small- and medium-scale enterprises, the relationship between personalization and efficiency is particularly significant, given the challenge of improving sustainability efficiency with cost considerations. Sustainability, according to Industry 5.0, is no longer a boundary but a by-product of responsive manufacturing systems.

Most importantly, however, hyper-personalization does not mean that SMEs need to become overly complex or technology-focused. When successfully implemented, the process typically begins with a small but variable list of options that can be expanded as technology improves. The goal is not complete customization but delivering meaningful options to customers while being able to keep operations in balance. The use of technology as a coordinator ensures that personalization can be handled without draining resources.

In the future of smart manufacturing, hyper-personalization and agile manufacturing shift us from creating the same thing for efficiency reasons to creating something that meets a customer's needs. Small- and Medium-Sized Enterprises (SMEs) use flexibility and human know-how as strengths because they are agile and have a human connection with customers that larger companies cannot compete with. Industry 5.0 technology enhances these; they do not diminish them. When technology and human quick thinking are combined, "SMEs can lead the future of manufacturing as one that focuses on relevance and connection rather than mere scale."

8.4 Generative AI and Autonomous Manufacturing Intelligence

In a manufacturing system increasingly complex and replete with data, the situation now in small- to medium-sized companies is less about transforming information into insight than finding the information in the first place. This is where generative AI technology revolutionizes the process of transformation. Conventional AI systems examine data for predictive outcomes, but with generative technology, it becomes possible to produce new solutions in the form of designs, plans, scenarios, or recommendations based on predefined parameters. This capacity in Industry 5.0 does not suppress human creativity; it enhances it.

Generative AI introduces a whole new paradigm of collaboration between humans and machines. Engineers, designers, and users are no longer only getting forecasts from AI. Rather, they are collaborating side by side with AI. A human has some aims to make something lighter, cheaper, more energy-efficient, or faster to transport, and the AI browses through numerous solutions that would be difficult to evaluate manually. Afterward, the human relies on their personal judgment and values to select and optimize the optimal solution among the options available.

For small- to medium-sized enterprises, the impact of generative AI is even more significant because of its ability to facilitate more advanced engineering or optimization. Earlier, aerospace engineers would need specialized staff and high-end software to perform tasks such as the optimization of complex designs, the simulation of multiple parameters within an engineering process, or even long-term capacity forecasting. With the availability of generative AI solutions and their easy-to-use interfaces, even a small industry would be able to apply these solutions to redesign the manufacture of an existing component or consider different manufacturing layouts as well as different schedules.

Generative AI in product development allows smaller companies to move past a slight improvement in their product designs to test novel exploratory designs. Many possible product designs are generated and tested by generative AI, presenting a person with alternatives they might not have conceived on their own. These designs will often be an improvement since they are able to balance factors better than in previous designs, such as strength versus weight in a product. The decision falls on a human. Generative AI recommends; humans choose.

Apart from design, generative AI also aids in the intelligence of autonomous manufacturing. Such systems keep learning and providing suggestions for improvement. For instance, an AI model considers data regarding manufacturing output, quality trends, energy consumption, as well as the supply chain to provide suggestions for improvement in terms of better configurations or policies. For instance, a generative model would provide suggestions regarding a different order of manufacturing or varying sizes of manufacturing with a focus on efficiency and reacting to changes quickly.

In an SME business environment, autonomy does not translate to losing control of the situation. It means having recommendations as fast as the machine can think. People have the freedom to decide upon receiving the recommendations. This aspect needs to be noted and remembered clearly. Autonomous systems are capable of functioning in standardized environments with low variability, while SME business environments require flexibility and decisions based on context. Basically, generative AI understands this and operates within boundaries defined through human values and goals.

Another major benefit of generative AI is the way in which it assists in the transfer of knowledge. Smaller organizations tend to rely on the local expertise that more senior staff have. As these employees retire or leave the company, a lot of the knowledge that has built up gets lost. The generative AI model learns from the decisions that have been made and the feedback provided, gradually building the expertise that can then be modeled for use in the future.

Ethical as well as governance issues arise when considering the application of generative AI technology in manufacturing. Industry 5.0 is all about openness, responsibility, and trust. The application of generative AI technology in small- and medium-scale enterprises must incorporate explainability, tracing, and human interventions or control over the entire process. If applied wisely, implementers gain greater trust as a result, as explanations become clear, and success is easily measurable.

In the end, however, the use of generative AI and autonomous manufacturing intelligence represents a different kind of capability with the potential to allow small- and medium-sized enterprises to think bigger, experiment more safely, and make better decisions. It enables a blend of computer creativity and human know-how, giving smaller firms access to strategic insights that are today the preserve of larger enterprises.

In the future smart factory, it will not be generative AI that authoritatively gives the orders. Instead, it will be operating in the background analyzing possibilities, providing insights, and assisting in decision-making with people in charge all the while. This complementary interplay between autonomy in AI and human control is precisely the promise of Industry 5.0, the future manufacturing system in which smart thinking flourishes, complexity is managed, and human creativity remains relevant.

8.5 Sustainable and Resilient Smart Factories Through AI

Sustainability is now mainstream in the world of manufacturing, rather than just a secondary consideration. Where medium- and small-sized businesses are concerned, the impact is twofold—with smaller profit margins and less to invest compared to large corporations, they are finding themselves increasingly pushed by their clients, the authorities, and suppliers to reduce their impact on the environment while proving their responsible credentials. This is where AI comes into the picture, offering medium- and small-sized businesses solutions to these needs without necessarily generating additional work.

At the factory level, the role of artificial intelligence is to make the manufacturing process more energy-aware. This is achieved through the continuous monitoring of how the energy is being consumed. The old way of energy conservation is estimated through the monthly bills received from the energy provider. This method does not show where the energy is being wasted. The new method of artificial intelligence relies on the energy usage information gathered from the sensors and the production plans. For small- and medium-sized enterprises, the reduced energy consumption translates to direct financial savings.

Material efficiency is another domain where AI makes a difference for sustainability. Waste in manufacturing refers to scrap, rework, and overflow inventory, which affects manufacturing as well as the environment negatively. AI techniques relate process parameters, machine states, and quality outcomes to identify waste-causing factors and determine why waste happens. Small- and medium-scale businesses reduce material use without curtailing quality because they eliminate waste-causing factors upstream, thus aligning sustainability with operating, instead of going against, a process.

Besides the issue of efficiency, AI also makes the system more resilient, which is a significant concept encompassing sustainability but not always highlighted. A resilient factory is one which can survive any form of disruption, whether a shortage of suppliers, varying demand patterns, poor weather conditions, or machinery failures, without wasting a lot and a significant number of days being idle. AI contributes to this concept in that it predicts the risk, simulates a scenario, and recommends flexibility. For example, where a supplier fails, AI would recommend alternative sources and methods to minimize delay and environmental degradation.

Industry 5.0 is centered on people for the purposes of sustainability. Smarter factories have to ensure they are healthy for the environment and the people who work for them. AI allows for the reduction of heavy and dangerous work, which is made easier by such technology, and makes work environments healthier and jobs last longer, which is crucial for smaller businesses where people wear multiple hats.

AI assists small and medium enterprises in circular manufacturing. It is a system where materials and products are reused, remanufactured, or recycled and not discarded. AI analytics follow how materials flow and how a product depreciates. It assists in the process of deciding whether a product has to be remanufactured or end-of-life processed or reused. It gives a clear understanding that helps SMEs extend the useful lives of materials and follow new regulations for circular economies with minimum administration.

Most importantly, a sustainable smart factory does not emerge as a result of revolutionary changes. It is a gradual process that is based on constant improvements. SMEs can begin by monitoring the usage on a few machines or monitoring the wastage at a single process point. They can then scale up as results emerge. They can start thinking differently by injecting AI perspectives into traditional SME operational methods. Thus, sustainability is integrated into their decision-making process.

In the future, product production will depend even more on managing production responsibly while things are uncertain. Smarter factories that apply AI for sustainability and resilience assist small and medium businesses in overcoming this hurdle while avoiding any impact on their speed and profitability. By balancing their concern for the planet, intelligence, and their designs in the context of people, SMEs can ensure smart factories that are both efficient and long-lasting, meaning they can provide value for many years.

8.6 Resilient Factories in an Era of Continuous Uncertainty

In today's manufacturing, what matters most and above all is not efficiency but unpredictability. Worldwide supply chain issues, wars and politics, pandemics, a lack of skilled labor, power variability, and climate disruptions have proven that traditional manufacturing models can be quite vulnerable. This poses almost insurmountable challenges to SMEs, which lack huge financial reserves to fall back upon, unlike larger companies. This makes business resilience not only an advisory but an imperative in strategy-building. Artificial intelligence emerges at this juncture to aid in creating factories that are less vulnerable to unpredictability but can perform effectively in an uncertain environment.

Ability to endure begins with understanding context. Problems balloon into crises when one overlooks signs or recognizes them too late. AI systems monitor such signs in the production flow, supply chain, logistics, and demand trends, identifying changes that can turn into disasters before they turn into failures. Smaller signs, such as increased supplier lead times, somewhat reduced efficiency in the machinery, or changes in customer orders, might appear unremarkable on their own. The power of AI is that it is able to connect the dots between the signs, allowing smaller to medium businesses to anticipate dangers before they have to simply react to threats.

In addition to problem identification, AI enhances resilience through scenario planning. Conventional planning relies on a forecast where everything remains constant. AI models perform numerous "what-if" analysis based on how various disruptions may impact the production process, the effects on inventory, the performance of deliveries, and the cost involved. For small- and medium-scale companies, AI brings planning out of the realm of guessing and instead helps with preparation. Business leaders can evaluate the available options, such as shifting operations to other plants and altering the number of batches or the use of alternative suppliers, before the occurrence of unavoidable events.

An intelligent factory has to be nimble in operations. AI assists in organizing employees based on work schedules and resource allocation that helps SMEs alter operations easily as changes occur. For instance, when a crucial machine is inaccessible, AI recommends different paths. When changes occur regarding employee availability, schedules are adjusted for a fair allocation of work without compromising on-time performance. These changes occur rapidly, but human oversight is always involved. Such a company is adaptable and has control. That is a virtue for SMEs that have narrow profit margins.

Human-centric resilience is an essential aspect of Industry 5.0. It is people who make a technology-based resilient system. AI assists in this by cutting the cognitive overload in times of crises. Instead of burdening managers with an enormous amount of data in times of crises, AI makes complexity simple and usable. This is especially true for small- and medium-scale businesses, in which only a few people are responsible for the entire operation. It can make the difference between success and failure.

Resilience also involves learning for an organization. Each crisis gives us a valuable lesson, but we tend to forget this when things go back to normal. The AI system gathers data from disruptions, including what actually transpired, what informed those decisions, and what were the consequences and uses this information for future guidance. In this way, over time, the factory increases its resilience, not by keeping disruptions at bay but by improving at coping with disruptions. Even small- and medium-scale businesses benefit from this process of continuous learning, inasmuch as organizational memory for past events takes a more precarious form.

Resilient factories are not hard, impenetrable fortresses. Instead, they are adaptive systems that change with circumstances. Artificial intelligence assists small- and medium-sized companies in their transformation from mere survivors to opportunity-seekers. While their competition stumbles through difficult times, resilient companies maintain their level of services, make their customers happy, and even increase their market share. In this manner, resiliency itself becomes a differentiating factor, rather than merely a tool for mitigating risks.

Resilience in Industry 5.0 is a combination of intelligent technology and human values. AI provides us with foresight, acceleration, and excellent analysis. Humans bring us what really counts—values, priorities, and ethics. By uniting these two aspects, we will have resilient manufacturing plants that are not just robust but nimble enough to recover from any setback without losing their way. For SMEs with a blurry outlook on the future, resilient manufacturing plants based on AI technology are not a choice—it's a starting point.

8.7 The Future SME Workforce: Skills, Roles, and Human–AI Co-Evolution

With the integration of smart manufacturing into Industry 5.0, what's taking center stage is not technology or software, but people. The future of manufacturing will not be about replacing people on the factory floor but about changing what people do inside smart systems. This is both an opportunity and a need that small- and medium-sized companies must adapt to because their workers are closely tied to daily operations.

In conventional factories, much work is task-oriented and reactive. This involves correcting equipment, rescheduling when there is a problem, and making decisions with compromised information. AI, which is now entering manufacturing and supply chains, is causing a paradigm shift where doing work is replaced by monitoring, understanding, and enhancing work. This is done by employees who are expected neither to know each and every task nor to respond to problems when they occur. Employees are working with intelligent systems that highlight insights, forecasts, and options. People's work involves moving upstream to prevent problems instead of correcting them.

This gives rise to hybrid jobs, an essential characteristic that defines the Industry 5.0 labor market. Instead of merely being a machine operator, one is now a process analyst who understands both the physical processes and the computer screens. Instead of being a maintenance technician, one now becomes a reliability engineer. Instead of a production planner, one is now a decision-maker comparing AI-generated scenarios with demands of customers and capabilities of the labor force. In SMEs, because the borders of jobs are more elastic, hybrid jobs emerge more effortlessly than in large organizations.

Skills shift with time. Where basic technical skills remain relevant, data literacy, systems thinking, and collaboration will also be necessary for employees working in SMEs. It does not mean, however, that all employees will have to become data scientists. Instead, they should feel confident when interpreting data, challenging AI recommendations, and leveraging technological support to inform decisions. Training shifts from a series of isolated courses to those that are embedded just as part of people's work routines. Bite-sized learning, digital support at work, and AI-enabled learning platforms will emerge as important factors in workforce development.

Industry 5.0 considers the well-being of the worker a strategic focus, not a simple procedure to adhere to. AI and automation minimize the risk associated with hazardous occupations, strain injuries, and mental exhaustion. Optimized schedules allow workloads to be distributed, and predictive analytics reduces the stress related to emergencies. For small- and medium-scale enterprises, the loss of a skilled employee can be detrimental; such innovations work to retain the workforce and keep the enterprise afloat. A healthier workforce not only performs better but also better withstands challenges.

The other crucial component of the future labor force is trust when collaborating on a task using AI. Small- and medium-scale businesses succeed when their employees trust the role of the AI system to assist them and not replace them. Transparency and letting them in on the process and decisions made are essential. Employees should understand the basis on which the AI determines its decisions and feel that they have the freedom to challenge the decisions made by the AI. This is achieved in Industry 5.0 because intelligence is shared and not centralized.

This is particularly important because of the demographics involved. With the older people in the workforce retiring and the entry into the workforce of the new generation, small- and medium-sized businesses find themselves at a knowledge gap and an opportunity to improve. This new generation is looking for innovation, engagement, and opportunities for growth. Learning, creativity, and collaboration-based smart manufacturing sounds far more attractive as an option rather than manual and purely automated alternative environments. Through Industry 5.0 thinking, SMEs are not just places people must work but places people want to work.

Ultimately, the future SME labor force will emerge through growth with technology. Humans and intelligent systems are learning alongside one another, adapting alongside one another, and enhancing one another. AI is equipped with speed, memory, and big-data capabilities. Humans are equipped with context, values, and meaning. This complementary dynamic makes production efficient as well as significant.

For small- and medium-sized companies, people-first strategies will not cost you; they will give you a competitive advantage. And companies that invest in their human resources and provide them with smart technology will outperform companies pursuing automation without considering the human aspect. In Industry 5.0, the best factory won't always be the one with the lowest number of human resources, but the one with the smartest human resources.

8.8 A Practical Roadmap to Future-Ready Smart Factories

The transition to Industry 5.0 does not mean that small and medium enterprises need to revamp all aspects of their business at once. The most effective transformations are achieved over time, and it's essential to have specific intentions related to what your business requires. The wise factory of tomorrow evolves incrementally, and to do this, one needs to first make things more visible, then add intelligent thinking, and finally improve human and machine collaboration.

The process begins with being viewed online. Small and medium enterprises can link key machines, gather credible data, and simplify processes with easy-to-view dashboards. Now that the data integrity issue is addressed, SMEs can leverage AI to provide predictive maintenance, schedule planning, inventory optimization, and demand sensing, which all provide rapid and straightforward value. As technologies advance, SMEs can engage and interact with machines through collaborative robots, generative AI, and human-centric interfaces that can enhance employee abilities.

The final stage is that of adaptive resilience. Smart factories are learning, adapting, and improving over time. AI assists in planning for various outcomes, risk analysis, and satisfaction of sustainability challenges, and humans remain in control and guide this process ethically. If SMEs implement this, they won't be miniature versions of giants but rather dynamic, intelligent, and people-centered establishments that are ready to compete and win in this Industry 5.0 period.

Zeitfracht Medien GmbH
Ferdinand-Jühlke-Straße 7
99095 Erfurt, Deutschland
produktsicherheit@kolibri360.de